Introduction to
Chemical Exposure
and
Risk Assessment

W. Brock Neely

 CRC Press
Taylor & Francis Group
Boca Raton London New York

CRC Press is an imprint of the
Taylor & Francis Group, an **informa** business

CRC Press
Taylor & Francis Group
6000 Broken Sound Parkway NW, Suite 300
Boca Raton, FL 33487-2742

First issued in paperback 2020

ISBN 13: 978-0-367-57976-0 (pbk)
ISBN 13: 978-1-56670-094-8 (hbk)

Visit the Taylor & Francis Web site at
http://www.taylorandfrancis.com

and the CRC Press Web site at
http://www.crcpress.com

Library of Congress Cataloging-in-Publication Data

Neely, W. Brock (Wesley Brock). 1926—
    Introduction to chemical exposure and risk assessment / by W. Brock Neely
        p.   cm.
    Includes bibliographical references and index.
    ISBN 1-56670-094-9
    1. Health risk assessment 2. Toxicology. 3. Environmental health. I. Title.
RA566.27N.44   1994
615.9'02—dc20
                                                                94-22452
                                                                CIP

Library of Congress Card Number 94-22452

PREFACE

This book was written to provide an introduction to the principles involved in assessing the risks from chemical exposure. It grew out of the author's interest in teaching students how to form their own opinions as opposed to relying on the popular press for answers to the problems faced from overexposure to chemicals. In teaching the course, the key elements were found to be distributed over many books. In addition, most of these texts were beyond the average student's ability to comprehend. It was felt that the subject need not be that complicated. From this frustration, a text outlining the principles used in making risk assessments was developed. The book is written for both the general reader who has an interest in the subject and the student taking a more formal course in the field of environmental science.

These principles include how risk is perceived (Chapter 1), how numbers are handled (Chapter 2), and how chemicals affect health (Chapters 3 and 4). From here the book describes the properties of the environmental sinks where chemicals are distributed (Chapters 5 and 6) and how environmental concentrations are estimated and matched with toxic effects to make an assessment of the risk (Chapter 7). The book concludes with a philosophical discussion of the important topic of risk benefit analysis (Chapter 8) and a presentation of several case studies (Chapter 9) in which the principles enunciated in the previous chapters are used.

Many people have read portions of the manuscript. Through their efforts the book was much improved. While these individuals were most helpful, full responsibility for all statements lies with the author. For laboring over the first attempt at writing about the legal aspects of risk benefit, I want to thank J.H. Gherlein, Charles Donnelly, and Robert Nault. The chapter on statistics was critically reviewed by Dr. Colin Park. The chapters on toxicology and cancer were read by Drs. Robert Moolenaar and David Neely. Dr. Richard Kociba critiqued the case study on dioxin. Finally, special thanks goes to Dr. Michael Kamrin who read and reread the manuscript, offering valuable criticism, much of which has been incorporated into the final version.

This book bridges the gap between popular articles and detailed scientific material. The text has been used in teaching classes ranging from liberal arts students to third-year students majoring in chemistry.

AUTHOR

W. Brock Neely, born and raised in London, Ontario, Canada, received his bachelor's degree in agricultural chemistry from the University of Toronto. From there he attended Michigan State University and earned his Ph.D. in biochemistry. After two years of postdoctoral training at Ohio State University and the University of Birmingham in England, he joined the Dow Chemical Company in Midland, MI.

He worked at Dow in various capacities from 1957 to 1986. In 1986 he formed EnviroSoft, a company devoted to writing computer software packages and providing consulting services to the chemical industry in environmental affairs. Throughout his career, his research interests have been in the area of analyzing the fate and movement of chemicals added to the environment. He has published several articles and books, and has presented papers at international scientific meetings. He has been a member of the EPA's Science Advisory Board and received the Synthetic Organic Manufacturer's Association gold medal for outstanding work in environmental science.

During this time Dr. Neely has had an interest in teaching science. This is illustrated with teaching assignments as adjunct professor at Johns Hopkins University and Michigan State University. More recently he has been affiliated with Saginaw State University, where most of the material in this book had its origins. A major interest of the author since 1986 has been teaching students how to make their own assessments of the risk from exposure to chemicals in the environment. The notes used in the lectures have been incorporated in the present text. The material has been used successfully in the teaching of risk assessment to both liberal arts and undergraduate science majors.

CONTENTS

1   RISK PERCEPTION

The headlines herald the message, "There is a Hole in the Sky." The story continues, "The loss of ozone in the Antarctic poses a global threat . . . If the trend continues, the resultant increase in skin cancer and other assorted ailments will be catastrophic." On a local level there are frequent articles suggesting a threat from cancer if too many fish are consumed from the nearby stream.

- Are we at risk?
- How much ozone can be lost before we have a catastrophe?
- How much fish consumption is too much?
- What is the risk and how do we assess the importance?

Remember, risk and fear are common elements in our lives, and usually go hand in hand. For example, fear drives us to shun risk—the greater the risk, the greater the fear. Those who fear flying will do anything to avoid the perceived risk, even if it means staying home. On the other hand, risk accompanies any action, no matter how trivial. To cite one extreme example, scientists at the University of Pittsburgh recently reported that taking a shower can be harmful. During a shower, toxic chemicals are released from the water, exposing the person to chemicals that may be up to 100 times greater than normal exposure. Is there a risk in showering? The simple answer is yes. Should individuals be afraid to shower? Definitely not, because the risk is too small to worry about. The problem is that scientists can measure things with greater precision and with more sophisticated equipment than ever before. Take a glass of water. Until 1970, scientists had managed to identify about 100 organic compounds in this glass of water. Now the number has grown to 500. It should come as no surprise that some of these newly discovered compounds are potentially harmful. Thus, in the case of the shower, while harmful chemicals can be found, the concentrations are so low that the risk is insignificant. There is a greater chance of slipping on the floor and having a fatal accident than suffering from exposure to any chemical that can be found in the water. So relax and enjoy the shower. People must learn—and most do—to balance the risks and thus steer a safe course through the potential minefield known as life!

In our daily lives, perceptions begin to achieve a certain credibility through sheer repetition. As an illustration, a program aired on "60 Minutes" by CBS on February 26, 1989, and repeated in the news media for several months talked about the use of Alar[1] on apples and the risk that people faced eating the contaminated fruit. The CBS documentary was very emotional:[2]

> Speaking against the background of a big red apple, emblazoned with a skull and crossbones, "60 Minutes" correspondent Ed Bradley intoned, "The most potent cancer-causing agent in our food supply is a substance sprayed on apples to keep them on the trees longer and make them look better. That's the conclusion of most scientific experts, and who is most at risk? Children who may some day develop cancer from being exposed to this product."

An estimated 50 million viewers of television saw the show. The program was based on the assumption that what these experts said was true. Mothers were frightened into dumping apple juice down the drain and school authorities halted the distribution of apples in school cafeterias. One goal of this book is to teach people how to analyze the perceptions that are created by such a media event and to provide the necessary analytical tools to assess the risk. The case study dealing with Alar (see Chapter 9) discusses this issue further.

Perception of risk is normally based on whatever knowledge is available. If the database is faulty, then the analysis of the risk will be faulty and the decisions made will be equally faulty. Two psychologists, Kahneman and Tversky, have been studying these and related matters for almost two decades (McKean, 1985). While many of their hypotheses are not universally accepted, the framework is being increasingly validated.

If one considers the U.S. population, which is the more likely cause of death and by how much, emphysema or cancer? Studies have shown that, when death rates between two causes are greater than two, respondents arrived at the right order but did poorly on guessing the ratios. For example, most groups correctly declared emphysema was less likely to cause death than cancer. They did poorly in estimating the true ratio (30:1). Table 1 gives the death rates for various causes of death (Bureau of the Census, 1982).

Tversky and Kahneman (Kahneman et al., 1982) have argued that, when judgments involving uncertainties are made, several rules known as heuristics[3] are adopted to help in the decision-making.

The first heuristic is known as *availability*. In essence this says that people assess the probability of an event based on their ability to think of previous occurrences (McKean, 1985). For example, is the letter *k* more often at the

---

[1] Alar is a registered trademark of Uniroyal Chemical Co. for the chemical known as daminozide.

[2] The following account was published March 18, 1989, by Accuracy in Media, Inc., 1275 K St. NW, Suite 1150, Washington, DC 20005.

[3] Heuristic: a rule of thumb that experience has shown to be a useful tool, but one without a firm theoretical or logical foundation.

Table 1. Leading Causes of Death in the U.S.

	Death rate in 1979	
Cause	Per 100,000	Per U.S. population
All causes	869	2,170,000
Cardiovascular	435	1,090,000
Cancer	183	458,000
Diabetes	15	37,500
Accidents		
Motor	24	60,000
Falls	6	15,000
Drowning	2.6	6,500
All other	15	37,500
Suicide	12.4	31,000
Homicide	10.2	25,500
Anemia	1.4	3,500
Viral hepatitis	0.2	500
Emphysema	6.2	15,500

beginning of a word or is it the third letter of the word? Based on familiarity, words that start with k are much easier to recall as opposed to words where the third letter is k. Consequently, the normal response to this question is that k is more apt to start a word. In fact k appears about twice as often as the third letter.

Another example involves the deadly botulism organism (Morgan, 1985). Death from botulism is a rare event, and consequently the occurrence becomes a newsworthy event. On the other hand, death by cancer is common and hardly ever reported. This leaves one with the false impression that botulism is more prevalent than it really is. Table 1 shows that every minute one person dies of cancer in the U.S. and death from botulism is so rare that the rate is not reported in this table. Subjects presented with the question dealing with the frequency of death from botulism and cancer systematically overestimated the frequency from the former, and underestimate the frequency of death from the latter. This is the same reason that safety studies of nuclear power plants are rarely convincing. Engineers try to imagine all the ways a plant can fail and then attempt to prove how unlikely the catastrophe will be. The very act of telling people about an event that has a low probability of occurrence makes it appear more probable and seemingly more likely.

The second heuristic suggested by Tversky and Kahneman (Kahneman et al., 1982) is known as *anchoring and adjustment*. The investigators suggested that people estimate the occurrence of an event based on their experiences. This creates a problem in that the estimate becomes an "anchor" which influences decisions made with respect to other events. The following experiment shows the effect of this heuristic. A group of subjects was told that 50,000 people die in one year in automobile accidents. The second group was told that 1000 people die each year from electrocution. Each group was asked to estimate the annual number of deaths from a variety of other causes. In both cases the ordering was correct. However, the influence of anchoring their judgment on the lower number caused the results to be shifted downward. In other words, the second group underestimated the number of deaths from causes such as cancer, because they were using a low number of deaths from electrocution as an anchor.

The final heuristic that needs to be discussed has been called *representativeness* (Kahneman and Tversky, 1973). This problem-solving heuristic is a shortcut the mind takes in dealing with an issue that is so complicated that it cannot be solved. In answering questions such as, "What is the probability that object A belongs to class B?", people tend to decide based on how much A represents or resembles B. An example is taken from an article in *Discovery* (McKean, 1985): Linda is 31, single, outspoken, and very bright. She majored in philosophy in college. As a student she was deeply concerned with discrimination and other social issues and participated in antinuclear demonstrations. Which statement is more likely?

1.   Linda is a bank teller
2.   Linda is a bank teller and active in the feminist movement.

In this example, people think Linda is more likely to be both a teller and a member of the feminist movement since feminist seems more representative of Linda than being a bank teller. However, a principle of probability is that the likelihood of two uncertain events happening together is always less than either event happening alone. Therefore, the likelihood that Linda is both a teller and a feminist is less than simply being a teller.

The heuristic of representativeness is a most important idea and there are many examples. One of my favorites deals with deer hunting. We are all aware of the "fact" that deer hunters tend to shoot each other in the annual fall trek to the woods. An article in one of Michigan's newspapers[4] analyzed the headline "Six Hunters Die." The headline leaves the casual reader with the thought that deer hunting continues to be a hazard to your health. The heuristic of representativeness suggests that readers are conditioned to the idea that deer hunting is associated with shooting hunters. On examining the article, Opre came up with the following analysis. Hunter No. 1 was murdered—an incident that is independent of deer. Hunter No. 2 wasn't shot but fell out of a tree; while tragic, it too was not dependent on deer. Hunter No. 3 died of a heart attack as did Hunter No. 4. Hunter No. 5 was found dead in a burning deer blind. The article did not say if the fire killed him, or

---

[4] An article by T. Opre in the *Detroit Free Press*, Nov. 20, 1988.

perhaps he also had a heart attack. Hunter No. 6 was found accidentally shot to death. Before reaching a conclusion that deer hunting in the woods is a dangerous sport, more study is needed. It is conceivable that a person may be safer in the woods during hunting season than on the streets of some of our major cities.

A further illustration of our distorted perceptions about risk is the fact that 50,000 deaths occur every year in traffic accidents, with half being attributed to drunk driving. This fact is accepted with remarkably little comment. On the other hand, the great tampon-toxic shock scare of a few years ago received a great deal of public attention, even though the total number of victims scarcely equaled one weekend's traffic fatalities. In a similar fashion, the yearly death toll related to cigarette smoking in the U.S. is over half a million. This point receives a modest amount of attention although the number of smoking-related deaths is equivalent to three fully loaded 747 passenger jets crashing every day. However, even a single jet crash becomes a major news item attracting worldwide coverage. For a brief period travel by airline is sharply reduced and the business of selling air travel insurance surges. Four general observations emerge from studying these types of paradoxes.

1. Sudden, surprising news triggers shock, horror, and fear.
2. People find greater misfortune in an accident that befalls a group than if the same number of individuals suffer the same kind of accident individually.
3. In voluntary activities, people will accept risk that is 10 to 100 times higher than from activities or circumstances that are imposed without their consent. You only have to visit a ski resort to see the voluntary risks that people experience—and then visit the hospital serving the resort to see some consequences.
4. There is a clear indication that people want to be informed about the nature and extent of risks that our complex world creates. When the public is informed in a straightforward manner, they will accept more risk than if either misinformed or ignored.

One of the root problems in public discussion of risk-related matters is that the plight of the victim is always emotionally gripping. This is the basic tenant of journalism: give most of the space to the victim and a few human interest notes about the survivors, but don't waste time on the nonvictim who walked the same path as the victim and survived. To understand risk, it is necessary to take a big step beyond the journalistic view to perceive the number of victims in relation to the number of people who did just what the victim did and managed to escape unscathed. This is exactly how insurance companies operate. They have no choice; they must collect premiums from the nonvictims to pay the claims of the victims, cover the operating costs of the company, and show a profit. Once these various factors influencing the perception of risk are understood, it becomes much easier to evaluate the risk and place it in perspective.

There are many types of risk ranging from crossing a busy street to the very complex issue represented by eating food produced with the aid of pesticides. In the latter case the consumer has a small risk from being exposed to trace amounts of chemicals while enjoying the benefits of a productive, low-cost food supply. On the other hand, the producer faces a bigger risk from being exposed to higher concentrations of the chemicals. Their compensation for facing this risk is a greater monetary reward from the more efficient production of food that occurs with the aid of pesticides. The issue is complex since the farmer voluntarily makes his decisions to face the risk while the consumer has no real choice and often is not even aware of the risk or the benefit. How these types of hazards are quantified and how society attempts to resolve the question of risk vs. benefit is the subject matter of this book.

Before proceeding, the expression *risk assessment* needs to be defined. There are many types of risk assessment such as evaluating the risk from falling down stairs, riding in an automobile, skiing, etc. This book will deal with analyzing the risk from exposure to chemicals that are found in the environment. The following is taken from a study by the National Academy of Sciences (NAS, 1983).

Environmental Risk Assessment is the characterization of potential adverse effects on humans exposed to chemicals in the environment. To describe the process in greater detail, the NAS identified four distinct steps in performing a risk assessment.

1. The first is the *identification of the hazard.* For example, does exposure to a chemical agent cause an increase in cancer or does a decrease in the ozone layer cause skin cancer?
2. The second step, once the hazard has been identified, is to establish the relation between the *dose of the agent* and the incidence of the adverse effect. These types of relationships are established using animals as the test species and then making the extrapolation to humans.
3. Once the dose response has been found, the third step visualizes an *exposure assessment.* How much exposure is there to the hazard? How does this exposure relate to the dose that was measured in step 2?
4. Finally, all of the above is integrated into a *risk characterization.* This process estimates the incidence of a health effect under the various conditions of the exposure assessment.

While the above was written specifically for the hazard to humans, the same principles can be applied to other systems such as aquatic organisms. How these four steps are reduced to practice constitutes the principles of risk

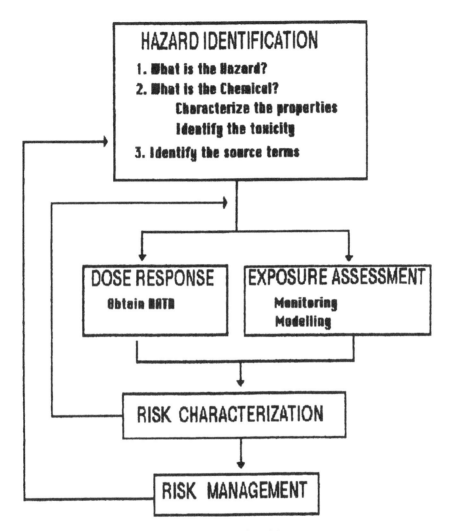

Figure 1.   **Diagram illustrating the risk assessment process.**

assessment. Figure 1 shows the process as a flow chart. The figure emphasizes the constant iteration that takes place in performing an assessment. In other words, the evaluation is a dynamic operation always feeding on new data until all parties are satisfied that the risk has been properly characterized. The remainder of the book will be involved in describing these principles in greater detail. A brief summary of the chapters follows.

Numbers are a very important part of risk assessment and it becomes essential that a clear understanding is made about their significance and how they are generated. Chapter 2 addresses this topic by discussing statistics and how this science is used to make a determination of the validity of an investigation and what elements make up the best experimental design. The significance of this latter subject will be more evident as the various case studies in Chapter 9 are discussed.

In this book emphasis will be placed on the risk to humans from exposure to chemicals. Thus, Chapters 3 and 4 describe the science of toxicology and how data are generated to establish the impact of a chemical on humans. It must be remembered that effects on other species besides humans can also be studied. For example, in the early 1960s the wide distribution of DDT was cited as evidence for a decrease in the population of certain birds (Carson, 1962). Thus, the main effect investigated was the toxic action of DDT on birds.

The third element of risk assessment deals with exposure. Chapters 5 and 6 cover the two main media by which organisms become exposed, air and water.

Chapter 7 deals with the quantitative aspects of exposure. By modeling the movement of chemicals in the environment, estimations of concentrations can be made. Using these estimations, a comparison with toxicological data will indicate the potential for risk.

Once the hazard and exposure are characterized, an analysis of the risk vs. the benefit must be made. How society presently deals with the complex problem of solving the risk benefit equation is discussed in Chapter 8.

Finally, case studies dealing with dioxin, radon, ozone, and Alar are presented in Chapter 9. These studies will show how the elements of Chapters 2 to 7 are integrated into a final package from which a decision can be made as to the risk and the benefit.

2   STATISTICS

INTRODUCTION

Interest in statistics is due in part to the recognition of the important place that working with numbers has in our everyday life. There are many examples where contact is made, such as opinion polls (particularly in an election year), national economic and social statistics, health and environmental data, etc. Comprehension can only come through a study of statistics. As the many examples suggest, data are not merely numbers but numbers with a context or meaning (Moore, 1990). The number 10 without context carries no information. However, the idea that a hammer costs $10 per unit allows comments to be made on the value of the hammer. As another example, calculating the mean of five numbers is an exercise in arithmetic, not statistics. However, calculating the mean price of a gallon of gasoline is statistics, particularly when combined with a look at the spread in the price from one location to another.

This chapter is intended to provide the student with a grasp of the rationale underlying the application of statistics to environmental problems. The task will be undertaken through a brief exploration of the following subjects.

1. Experimental design
2. Sampling
3. Distributions
4. Confidence intervals
5. Units
6. Conclusion

Discussions on probability, expected value, and decision rules will be found in Appendix I. While these three topics are important elements in statistics, they are not essential for an understanding of the material in this book.

EXPERIMENTAL DESIGN

One important use of statistics is in the interpretation of experimental results. Thus, when the newspapers report that eating fish caught from a river will cause cancer, the facts need to be examined before assessing the risk. The first step in such an examination is the study of the experiment on which the conclusions were based. By understanding and applying the principles of experimental design, a decision can be made regarding the validity of the study under question. The problem of design deals with how to plan the experiment to answer specific questions. For example, to decide the carcinogenicity of a chemical, animals are exposed and the initiation of tumors is observed. Such deliberate interventions are known as treatments. A result, such as the initiation of tumors produced by a treatment, is known as an effect. These two elements, treatment and effect, are the basis for the discussion that follows. Experiments and results as reported by the media should be carefully studied before their conclusions are accepted. For example, a film strip on the television news channel shows a spray emerging from an airplane flying low over a field. The reporter states that the chemical in the spray is drifting away from the field and several farm animals are exposed. The next segment of film shows some deformed animals and the conclusion is drawn that the deformities are a result of the chemical in the spray. In this example the reporter is claiming that the treatment (exposure of animals to the spray) caused the effect (deformities). Before accepting this conclusion, the details of the report need to be examined.

1. What was the magnitude of the exposure? All that is known is that spray drifted away from the target field and contacted some animals. The first critical step in any experiment is to establish a dose-response relation. Obviously, in the reported incident there is no way of knowing the degree of chemical exposure.

2. Contact with a spray containing an unknown amount of chemical was alleged to cause the observed deformities. Before such a conclusion can be reached, the previous history of the animals should be established. Could they have been exposed to other agents that might have caused the effect or were they even exposed to the chemical in question? This is a classic case of an uncontrolled experiment. At the very minimum, two sets of animals with identical previous histories are required. One group would be exposed to the spray containing the chemical, while the control group would be treated identically except that the spray would not have contained any chemical. Without adequate controls there is no way of determining if the deformities were caused by the drift or by some other event not reported, or were part of the natural background incidence. A main ingredient in a good experimental

design is the inclusion of a concurrent control. These are controls where the subjects are treated identically in all aspects other than the treatment. There may be times when the experiment cannot include such controls. However good the reason may be, the conclusions are bound to be clouded by some degree of uncertainty. A common illustration of the experiment where there is good reason for the lack of concurrent controls is the "before-after" comparison. A certain drug is generally believed to prolong the survival of children suffering from leukemia. It is true that the drug-treated children of today survive many months longer than did victims of the disease before the introduction of the drug. During the same period, however, many other advances in medical care have occurred. Would today's patients do as poorly without the drug as did their counterparts several years ago? The question can no longer be answered experimentally, because it would not be ethical to establish a concurrent control group of patients by withholding a drug that is believed to be effective. This peculiar difficulty in human experimentation points to the importance of conducting thorough conclusive experiments early in the trial period of a new drug.

Another important item to be considered in establishing an experiment is to decide if the individuals were selected at random.[1] This is to ensure that all individual differences are randomly distributed between the control and the treated group; otherwise, differences in the outcome may really reflect innate differences among the subjects themselves. This subject is fraught with many difficulties. For example, the apparently simple task of dividing 50 mice into 2 equal groups for control and treatment can be a tricky operation. Removing 25 mice to another cage may result in selecting 25 mice that are more lethargic, therefore easier to catch. Groups of animals separated in this manner could not be used for any valid experiment. Whatever the outcome, observed results would be in doubt due to the characteristic of lethargic vs. active mice. The two groups need to be selected on a random basis. The key requirement is that no characteristic of a subject should play a part in the assignment to a group. Once random assignments have been made, the remaining concern is to subject all groups to identical conditions during the experiment, except the treatment under study. While this sounds easy and straightforward, there are many pitfalls. Treated and control mice in a labora-

---

[1] The subject of randomness needs to be understood. Random behavior refers to events that have uncertain individual outcomes but a regular pattern of outcomes in many repetitions (Moore, 1990). For example, the tossing of a coin has an unpredictable individual outcome (i.e., either a head or a tail); however, in the long term the proportion of heads to tails will be 50:50. Thus, when individuals are selected for an experiment, it is necessary to ensure that each person has an equal chance of being selected and that no bias in the selection process has been shown.

Table 1.   **Sample Size to Detect Various Differences at the 95%**
Confidence Level

Difference (%)	Animals per group
2	1237
5	309
8	137
12	77
16	49
20	34

tory situation must be moved on a regular basis to ensure that all animals are exposed to the same environment. This will prevent a situation where control or treated cages are subjected to either different temperature or light regions in the laboratory, i.e., one section may have a draft from the temperature control system used in the cage room. If the treated group received an injection, then the control group must be injected with a placebo.[2] Remember, controls must be identical in all respects other than the treatment. Finally, precautions need to be taken to rule out the natural bias of the investigator. The essence of this procedure is that the personnel who carry out the treatment should not know which subjects belong to which group. Only by careful attention to all details of the experimental design can valid conclusions be reached. Thus, the original illustration lacked all three elements. There was a lack of control, the subjects were not chosen at random, and the investigative reporter had a bias in reporting the results.

The final task in the experimental design is to select the correct number of animals to decide if a significant difference exists between a control and a treated group. For example, in planning an experiment to show the effect of a chemical carcinogen, a group of sufficient size must be selected to learn if a significant difference exists. Intuitively, it is realized that as the difference becomes smaller, the group of animals tested becomes larger. Exactly how large requires a greater knowledge of statistics than will be covered in this book. Table 1 lists the necessary sample size[3] to detect various differences at the 95% confidence level.[4]

Thus, for a chemical where the expected difference was 5%, the experiment would have to include 309 animals at each level of treatment including the control group. Obviously, as more animals are tested, the experiment becomes more costly. Consequently, the detection of small effects becomes very expensive.

---

[2] Placebo: a pill, preparation, etc. given as medicine, but actually containing no active ingredients.
[3] This table wes prepared by Dr. C. Park, a statistician with the Dow Chemical Co., Midland, MI.
[4] Confidence levels will be described more fully later.

SAMPLING

Once the experiment has been designed, the next step is to take samples and make a determination of the effect. From these samples an attempt is made to draw conclusions that will represent the total population. If the experiment was to decide the average weight of all students, there would be two ways to proceed.[5] Obviously, all students could be weighed and an average found. In this case the total class of students would be called the population and the average weight would be called the parameter. There would be no uncertainty in the result since all students were included. The second approach, and the one usually followed, is to infer from a sample of the students the average weight of the entire class.[6] In this case the average of the sample is called a statistic and used as an estimator of the corresponding population parameter. This is the reason that randomization of samples is a key element in experimental design. Only random samples will represent the total class or population. In such a case the resulting statistics will be fair estimators of the population parameters. Suppose that the true mean weight of all male[7] students is 165 lb. Several groups of students are chosen and the average weight in each group is computed. Comparison of these sample averages with the true mean indicates how well the sample statistic represents the population parameter. If the sample size is one, then the weight of each student becomes an unbiased estimate of the population mean. The individual weight is just as likely to be high as low, and the long-term average will approach the true mean. For larger group sizes, the mean weight in each sample is also an unbiased estimate of the population mean. Such samples are almost certain to contain weights above and below 165; therefore, the mean of the sample is likely to be closer to 165 than the weight of a randomly selected individual student. Obviously, the statistic comes closer to the parameter as the sample size increases. At the extreme, if the sample approaches the size of the population, the sample statistic is indistinguishable from the population parameter. Besides the sample size, the amount of variability in a population determines how accurately a statistic will estimate the corresponding parameter. For example, if no student's weight differed by more than a pound or two from the mean of 165, then even statistics from very small samples will be excellent estimators of the true mean.

Sampling can be illustrated with a bag containing 100 white and 100 red beads. By pulling a sample of 10, a decision has to be made regarding the exact proportion of white to red. It can be shown that there is a 25% chance

---

[5] The discussion that follows would apply equally as well to other effects such as the background incidence of cancer in a population of animals.

[6] In actual practice, several groups of students will be weighed and the average of each group combined to yield the estimated average for the entire class. The more samples that can be taken, the better the estimate.

[7] If all students regardless of sex are chosen, then attention must be paid to choosing the sex in a random fashion; otherwise, the sample mean will be biased by the lower weight of females.

Table 2.   Distribution of Red and White Beads in a Sample of 10
           from a Population of 100 White and 100 Red Beads

Number		
Red	White	Probability[a]
1	9	0.01
2	8	0.04 tail of curve
3	7	0.12
4	6	0.21
5	5	0.25
6	4	0.21
7	3	0.12
8	2	0.04 tail of curve
9	1	0.01

[a]Probabilities are on a scale of 0 (no chance of the event occurring) and 1 (100% occurrence of the event).

of pulling 5 red and 5 white. The chances of other ratios are shown in Table 2. This demonstration introduces the notion of decision-making and hypothesis testing.[8] Suppose that in the case of the beads, the hypothesis is that the bag contains an equal ratio of red and white beads (i.e., 100 red and 100 white). If the hypothesis is true, the ratio of red to white in a sample will fluctuate above and below 1:1. For example, the probability of drawing a sample containing fewer than 3 or more than 7 red beads out of 10 is the sum of the tails of the curve, i.e., 0.10 (0.04 + 0.01 + 0.01 + 0.04). If the sample does contain 3 red and 7 white beads, a decision has to be made regarding the ratio in the population. Based upon a sample size of 10, how often will it be correct to accept the hypothesis (a ratio of 1:1 for the red and white beads) and how often will the decision be in error? If the hypothesis of equal proportions is accepted whenever any sample contains between 3 and 7 red beads, then the level of significance is said to be 0.1 (i.e., the sum of the two tails in Table 1). This implies that in 10 times out of every 100 trials a sample containing fewer than 3 or greater than 7 red beads would occur, leading to the rejection of the hypothesis, when in fact the sample might have been a member of the original population, i.e., a population containing equal numbers of red and white beads.

Instead of beads it is instructive to examine a biological experiment dealing with mortality. Suppose a certain disease causes 50% mortality, i.e., eventually half the victims die. In a random group of 10 patients, the odds that all will die are approximately 1 in 1000 (so small it cannot be fit to the histogram in Table 2. The probability of observing fewer than 3 deaths is 0.05 (0.04 for 2

[8] Appendix I contains a more complete discussion on the topics of decision-making and hypothesis testing.

Table 3.   Distribution of Weights among 300 Students

Number of students	Weight (pounds)
1	A 130–135
1	B 135–140
2	C 140–145
18	D 145–150
21	E 150–155
44	F 155–160
66	G 160–165
60	H 165–170
40	I 170–175
25	J 175–180
10	K 180–185
7	L 185–190
4	M 190–195
1	N 195–200

deaths and 0.01 for 1 death). Ten patients are treated and the hypothesis is that the population mortality is 50%, i.e., no effect due to the treatment. The decision rule is if 3 or more patients die, accept the hypothesis. If fewer than 3 die, reject the hypothesis and accept the alternative hypothesis. In other words, the drug has reduced the mortality. Table 2 illustrates that in about 5 times out of 100, a conclusion would be drawn that the drug is effective while in reality it is worthless. Stated differently, there is a 5% chance that less than 3 patients would die if the sample represented the original population where the mortality was 50%. Setting of decision rules is discussed further in Appendix I.

DISTRIBUTION

The notion of distribution has been partially introduced through the discussion of sampling. Recall that the object was to learn the mean weight of all male students. To do this, a sample was taken and the weights were measured. As more measurements were made, the trend approached the population parameter of 165 pounds. It can also be visualized that there will be a few weights on either side of the mean representing extremes in over- and underweight. A possible tabulation of the results might resemble the data in Table 3. As more individuals are weighed, the distribution tends to smooth out and approach the continuous curve as shown in Figure 1. This theoretical curve toward which finite measurements often converge is known as the Gaussian curve, or simply the normal distribution.

Thus, if the task was to measure the length of a room, and the job was delegated to 100 people and the results tabulated and plotted, a distribution

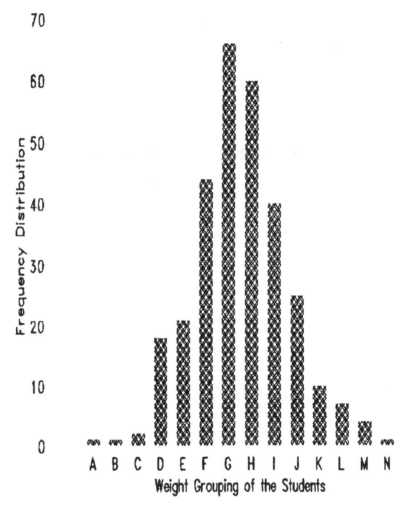

**Figure 1.   Data from Table 3 plotted as a distribution of weight vs. fre-
quency.**

curve similar to Figure 1 would result. The correct length would be represented
by the mean of all the measurements. It is the expected value or the long-
term arithmetic average.[9] The one other value that characterizes a normal
curve is the variance (σ^2), which describes the amount of spread (dispersion)
in the data, which in turn determines the breadth of the curve. In other words,
the variance represents how carefully the 100 people measured the length of
the room. The more care used in making the measurements would result in
a small variance and a narrow dispersion curve. Mathematically, the population

---

[9] Appendix I contains more discussion dealing with expected value.

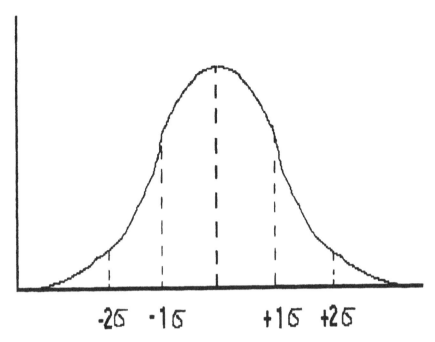

Figure 2. Normal distribution curve.

variance is the sum of the squares of the difference between the measurement (x) and the mean (μ) divided by the number of measurements (n). The reason for squaring ($x_i - \mu$) is to cancel the effect of the minus sign, i.e., some values of x_i are greater and some are smaller than the mean value. The square root of the variance is known as the standard deviation (σ) and is the distance along the x axis from the mean value (μ) to the point of inflection of the curve at the steepest point. Since it has the same units as the mean and represents an actual distance, the standard deviation is readily visualized (Figure 2).

In most applications, data for the entire population are not known. What is available is a sample. Using the sample, the standard deviation is measured to infer the deviation of the entire population. To evaluate σ for a sample the formula described above is adjusted as shown below. The use of (n − 1) compensates for the fact that only a small group is used to represent the population.

$$\frac{\Sigma(x_i - \mu)^2}{(n - 1)} \qquad (i = 1, 2, \ldots . n)$$

If the data are very variable, the population will contain a high proportion of values that deviate from the mean, the variance and standard deviation will

Table 4.   Area to the Left of z under the Normal Distribution Curve

z	Area
−4	0.00003
−3	0.001
−2.58	0.005
−2.33	0.01
−2	0.023
−1.96	0.025
−1.65	0.05
−1	0.16
0	0.5
1	0.84
1.65	0.95
1.96	0.975
2	0.977
2.33	0.99
2.58	0.995
3	0.999
4	0.99997

be large, and the distribution curve will be broad. The breadth of the curve is not determined by the absolute magnitude of σ but by its magnitude relative to μ. This is expressed as the coefficient of variation (CV), which is the ratio of σ to μ expressed as a percentage ($100 \times \sigma/\mu$). The CV is especially useful for comparing variability in different populations where the means may differ widely or whose data may be measured in entirely different units. For example, if two groups of students measured the length of the room, using feet and meters, the respective CV could be compared to learn which group had the tightest data.

Besides the standard deviation discussed above, another important characteristic of the normal distribution is the area under the curve. The total area represents the entire population of events, which is equivalent to absolute certainty. Accordingly, the area to the left of some known value, termed the z value, represents a fraction of the total population, or the probability of the event. To simplify the calculation of these areas, tables have been designed ranging from -4σ to $+4\sigma$.[10] For purposes of illustration an abbreviated list is shown in Table 4. To find the area to the left of a selected value, the table is examined. For z = 1, the area is 0.84. A few interpretations of this number are as follows:

---

[10] While the range of scores is unlimited, a score of 3 represents 999 out of a total of 1000 events. Similarly, a score of −3 represents only 1 out of a total of 1000 events.

1. 84% of the members of the normal population described have a z value of less than 1.
2. A member of the normal population with a value of 1 also has a percentile rank of 84.
3. The probability of randomly selecting a member from this normal population with a z = 1 or less is 0.84.

For most problems the investigator is dealing with raw scores. Such data may be converted to z values as shown below.

$$z = (X - \mu)/\sigma$$

where

μ = the mean
σ = the standard deviation
X = raw score

As an example, the weight of a population of rats was normally distributed with a μ of 14 oz. and a σ of 2 oz. If one of the rats weighed 12 oz., what is the percentile rank for this weight?
Solution:

1. z = (12 − 14)/2 = −1
2. From Table 3 the area to the left of z = −1 is 0.16.
3. The percentile rank for this rat is 16%, i.e., 16% of the rats will weigh less than 10 oz. and 84% weigh more than 10 oz.

CONFIDENCE INTERVALS

One important use of the mean and standard deviation is to decide the degree of confidence that can be placed on a measurement. To illustrate, several years ago a hypothesis was set forth (Ratcliffe, 1967) which stated that as the concentration of DDT (a synthetic pesticide) in the environment increased, there was a corresponding decrease in the thickness of eggshells from birds that had been exposed to the chemical. Thinner eggshells would result in an increase in the number of eggs broken in the nest and was one reason advanced for the observed decline in the reproduction rate of several bird species (Carson, 1962). The hypothesis was subjected to direct experimentation. This was accomplished by incorporating DDT into the diet of birds and measuring the eggshell thickness from the treated and the control animals (Bitman et al., 1969). In this case, the controls were treated identically in all respects except that the pesticide was not incorporated into the feed. The

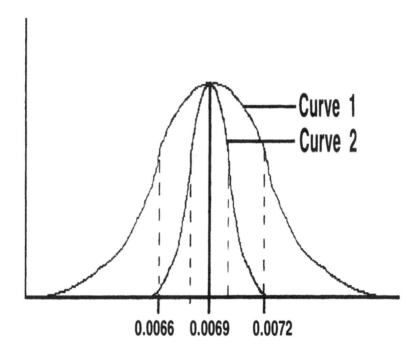

0.0066   0.0069   0.0072

Eggshell Thickness (inches)

Figure 3.   **Curve 1: normal distribution curve for the thickness of egg-shells with a standard deviation (σ) of 0.0003 inches. Curve 2: eggshells with a σ of 0.00003 inches.**

average shell thickness for a control sample of 368 eggs was reported to be 0.0069 inches. What should have been reported, and was not, was the standard deviation of the result. Without this statistic there is no way of knowing how close the individual measurements were to the mean value of 0.0069 inches. For the sake of the present discussion it will be assumed that the standard deviation was 0.0003 inches for a CV of 4.35%. Recall that 68% of the area under the normal curve is located between $\mu \pm \sigma$ or 0.0068 $\pm$ 0.0003. A normal distribution curve can be drawn as shown in Figure 3 (curve 1). For the eggs from the treated group the data suggested that 314 eggshells had a mean thickness of 0.0066 inches (Bitman et al., 1969). Could this sample have come from the control population or is it indeed a member of a population representing birds whose shell thickness has been affected by DDT? The answer is dependent upon the standard deviation. If the control population had the distribution shown in Figure 3 (curve 1), then obviously the treated sample could just as well have been a member of this group and the conclusion would have to be reached that treatment with DDT does not affect the thickness of eggshells. On the other hand, if the standard deviation of the control was 0.00003 so that the normal distribution looked like curve 2 in Figure 3, a different conclusion could have been reached. The treated sample with a mean of 0.0066 and a standard deviation similar to the control (0.00003) must be from a new population and the statement could be made with confidence that the treatment had an effect.[11]

UNITS

As mentioned at the start of this chapter, numbers by themselves are meaningless. To complete this introduction, a brief discussion of the importance of units is necessary. This can be a very confusing topic, especially for people living in the U.S., where a mix of English and metric systems are coming into display with greater and greater frequency. For example, as a person travels down the highway it is possible to see signs that give distances in both miles and kilometers. In Canada and Mexico, all the units are in metric. Some food packages are now reported in both grams and pounds. In this book the "SI" or "Système International d'Unités" (International System of Units) introduced in 1960 will be used, commonly referred to as metric units. These are shown in Table 5 with their English equivalents and the conversion factors.

There are several prefixes used to express the large and the small compared with a basic unit (Table 6). These precede the unit; thus, a microgram is 10^{-6} of a gram. The notation of 1 microgram (μg) is much easier to read and write than the equivalent 0.000001 gram.

One major difficulty with the SI system is the lack of perception or visualization of the magnitude of the units listed in Table 5. The following examples will illustrate what is meant.

1. For a person raised on the English system, a rate of 5.3 kg/ha means nothing; however, the equivalent rate of 5 pounds/acre may be readily visualized as 5 pounds spread over an area the size of a football field. What is lacking for most Americans is a mental picture of a kilogram and hectare.
2. A gram is a very small amount, being the equivalent of a few grains of salt or the amount of artificial sweetener in a package designed for a cup of coffee.
3. The barrel is a convenient unit and there is nothing comparable in the metric system. A cubic meter (m^3) contains slightly more than 6 barrels. The SI equivalent to the barrel is 158 liters.

---

[11] As indicated in Appendix I in the section dealing with decision rules, the Bitman experiments on eggshell thinning were reported to be highly significant by another statistical test.

Table 5.   SI Units and English Equivalents

Unit	English	SI	Conversion from English to SI (multiply by)
Length	Yard	Meter (m)	0.914
Area	Acre	Square meter (m²)	4040
	Acre	Hectare (ha)	2.47
		Square km (km)²	
Volume	Gallon[a]	Cubic meter (m³)	0.00378
	Cubic foot	m³	0.0283
	Barrel[a]	m³	0.159
Mass	Pound	Kilogram (kg)	0.453
	Ounce	Gram (g)	28.3
	Ton	Megagram (metric ton)[b]	907
Pressure	Atmosphere	Pascal (Pa)	101,300
	mm Hg (torr)[c]	Pa	133
Energy	Calorie	Joule (J)	4.184
	BTU	J	1,055
Temperature	Fahrenheit	Celsius	$(F - 32) \times 5/9$

[a]U.S. gallon and barrel. There are 42 gallons in one barrel.
[b]There are a million grams (Mg) in a metric ton (t), i.e., 1000 kilograms.
[c]While still found in the literature, these two equivalent terms (mm Hg and torr) should be considered obsolete.

A second difficulty arises from the prefixes in Table 6. Similar to the units themselves the prefixes do not convey a notion of size. Thus, the only difference between 1 pg (picogram) and 1 Pg (petagram) is that the p is capitalized in one and not the other. The idea that 1 Pg is 10^{27} (1 followed by 27 zeros) times as big as 1 pg is not conveyed. In spite of these difficulties there will be less confusion if the SI system is adopted and an effort made to develop the mental pictures necessary to use them intelligently.

Finally, special attention will be given to describing the units of concentration and pressure, two very important concepts in dealing with environmental processes. The units must be chosen with care. In this text the following system will be used:

1. Solution concentrations will be reported as g/m³ or g/L where a liter is equal to 0.001 m³ or approximately 1 qt. The gram may be preceded by one of the prefixes shown in Table 6.
2. Toxicological dosages will be cited as g/kg.

Table 6.   Common Prefixes to be Used for SI Units

Factor	Prefix	Factor	Prefix
10^1	Deka (da)	10^{-1}	Deci (d)
10^2	Hecto (h)	10^{-2}	Centi (c)
10^3	Kilo (k)	10^{-3}	Milli (m)
10^6	Mega (M)	10^{-6}	Micro (μ)
10^9	Giga (G)	10^{-9}	Nano (n)
10^{12}	Tera (T)	10^{-12}	Pico (p)
10^{15}	Peta (P)	10^{-15}	Femto (f)
10^{18}	Exa (E)	10^{-18}	Atto (a)

3. Air concentrations are routinely reported on a mol/mol basis.[12] This is a convenient method as it cancels the effect of declining air density with altitude. Thus, a trace gas such as CO_2 measured at 350 μmol/mol[13] will be the same at sea level or at the top of the troposphere. There are times when air concentrations are reported as g/m³ to place them on a comparable basis with water concentrations. To make the conversion from mol/mol to g/m³ it must be recalled that 1 mol of gas occupies 22.4×10^{-3} m³ at 0°C and 101 kPa (1 atmosphere).[14] Thus, by using the formula shown below, the conversion can be made.

$$\frac{g}{m^3} = \frac{air\ concentration\ (mol/mol) \times mol\ wt\ (g/mol)}{22.4 \times 10^{-3}\ m^3/mol}$$

For a concentration of 350 μmol CO_2/mol air the conversion yields:

$$\frac{g}{m^3} = \frac{350 \times 10^{-6}\ mol\ CO_2/mol\ air \times 44.01\ g/mol}{22.4 \times 10^{-3}\ m^3/mol}$$

$$= 0.7\ g/m^3$$

---

[12] A mole abbreviated as mol is similar in meaning to a dozen, thus a dozen apples contains 12 apples; so, a mole of any chemical contains 6.022×10^{23} atoms. Furthermore, the mass of a mole is simply the sum of the atomic masses. For example, the molecular weight of carbon dioxide (CO_2) is the mass of carbon (12.01) plus twice the mass of oxygen (16.00) or 44.01 g/mol.

[13] Note that 350 μmol/mol is equal to 350 ppm. Such units as ppt (parts per thousand or parts per trillion?), ppm (parts per million), and ppb (parts per billion) have found their way into the popular press. While they do convey some notion of size they can be very confusing and should be avoided. For example, a billion is 10^9 in North America and 10^{12} in Europe. Thus, when a concentration is reported as 2 ppb, what exactly is meant? Furthermore, confusion arises in comparing air concentrations of ppm which are on a mol/mol or volume/volume with water concentrations of ppm which are on a mass/volume basis. By using g/m³ or mol/mol there is no confusion.

[14] These conditions are called Standard Temperature and Pressure (STP), and are the conditions that exist at sea level. In reporting air concentrations on a g/m³ basis, it is always assumed that the measurement is made under STP conditions.

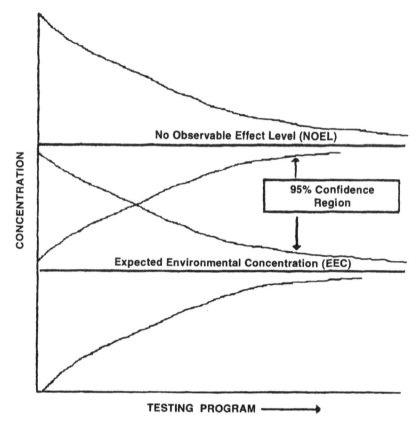

Figure 4. Schematic representation of how testing proceeds until the confidence regions indicate that the degree of uncertainty is reduced to the point where an assessment of the ratio can be made.

Recall that this is the concentration at sea level and 0°C. The value will be different for other altitudes and temperatures.

4.   Pressure is the force exerted per unit area of surface where force equals mass (kg) times acceleration (m/s²). Therefore:

$$\text{Pressure} = (\text{kg.m} / \text{s}^2) / \text{m}^2 \text{ or } \text{kg} / \text{m.s}^2 \text{ (Pascal)}$$

The pascal as defined is an extremely small unit. For example, the pressure exerted by a penny on the top of a table is about 100 Pa while the pressure exerted by 1 atm is 101,300 Pa. For convenience the kPa is commonly used; thus, 1 atm is 101.3 kPa.

CONCLUSION

The reason for this background is that these ideas play an important role in how society deals with the numbers used in addressing some of the environmental issues of the day. From the first chapter it will be recalled that the National Academy of Science (NAS, 1983) characterized four elements of risk assessment. The first was identifying the hazard followed by establishing the dose response for the effect. The third was deciding how much exposure there was to the hazard and finally integrating all of the above into a risk characterization.

These elements in risk assessment may be visualized by means of Figure 4. This diagram illustrates how testing and evaluation go hand in hand until the analyst is comfortable that the risk is acceptable. As the figure illustrates, the testing proceeds until the degree of uncertainty is reduced to the point where a definite statement can be made about the ratio between the expected environmental concentration (EEC) and the no observable effect level (NOEL). The next several chapters deal with toxicological, chemical, and environmental effects and how this ratio is estimated.

3　MAMMALIAN TOXICOLOGY

INTRODUCTION

Early in the history of civilization, humans in their quest for food must have attempted to eat a variety of materials. Through trial and error they discovered that certain substances produced varying degrees of illness up to and including death. From these observations two classes were identified, safe and harmful. The word "poison" was eventually used to designate all those substances that were distinctly harmful and "food" was used to indicate materials that were beneficial and necessary for body function. This concept of dividing chemicals into two categories has persisted to the present day. It is a useful idea as harmful materials can be readily placed into a category that is accorded due respect. In a scientific sense, such a strict classification is not warranted. Today drawing a line between beneficial and harmful substances is not possible.[1] Recognizing degrees of harm and safety is a more acceptable approach. Even the most innocuous of substances, when taken into the body in sufficient amounts, may lead to an undesirable response. In contrast, the most harmful of chemical agents if ingested in sufficiently small amounts will have no observable effects. From this analysis the harm or safety of a compound is related primarily to the amount of the chemical that is ingested.

To compare the effects of chemicals on animals of different size the amount is expressed as mass per unit weight of the animal. This is termed the dose. Thus, a gram of chemical fed to a 2-kg rat is a dose of 1000 mg/2 kg or 500 mg/kg. To equate this to a 70-kg person the dose would be the same but the amount fed would be 70 kg $\times$ 500 mg/kg or 35,000 mg (35 g). If a sufficient dose of any chemical is taken, some type of effect will be observed. As the dose is increased from a minimum to a maximum level, a graded response is observed rather than a sharp demarcation from no response to the ultimate response. A fundamental observation in the field of toxicology is the relationship that exists between the dose and the response of a chemical.

---

[1] However, the press and general public would like to have a clear separation between these two categories.

Thus, toxicology has developed into the study of the quantitative effects of chemicals on biological tissue. The discussion of this topic will be divided into the following sections and is taken from the book *Essentials of Toxicology* (Loomis, 1978).

1. Numbers in toxicology
2. Biological factors in toxicology
3. Chemical factors in toxicology
4. Route of administration
5. Testing procedures
6. Types of toxic effects

NUMBERS IN TOXICOLOGY

No chemical agent is entirely safe and no agent is entirely harmful. This idea is based on the premise that any chemical capable of causing a biological response will be inactive when the concentration or dose is below a minimal effective level. Consequently, there must be a range of concentrations that give a graded response between the two extremes of no observable effect and 100% response. The experimental determination of this range is the basis of a dose response study.

Dose Response Relationship

When a group of experimental animals is examined, differences between individual members of what is normally considered a homogeneous population are seldom obvious. Differences become evident only when the group of animals is challenged by a chemical exposure or some other type of treatment.

For example, inbred mice (similar to the common laboratory white mice) considered homogeneous are subjected to selected concentrations of a chemical. If the chemical can produce an effect, then a dose can be found to give that result. Individuals exposed to this particular level will respond differently. There will be a few animals that respond quickly while others will have no response. Thus, what was considered to be all or none applies only to a single member and when viewed for the entire group becomes a graded response. Such a result is due to the natural biological variation that is present in any population no matter how uniform. By plotting the dose against the number of animals responding, a dose response curve is obtained. For example, if several groups of mice were given various doses of chemical and the number of animals that responded were recorded, the results might look like those in Table 1.

Table 1. Hypothetical Data Showing the Relation Between Dose and the Number of Animals Responding

Dose (mg/kg)	No. of animals responding
5	0
10	1
20	1
25	3
30	5
40	9
50	10

In this experiment each dose level represents a group of mice. Thus, 7 groups of mice (10 in each group) are required to obtain the above results. The initial dose of 5 mg/kg is so small that no effect is seen; on the other hand at the high dose (50 mg/kg) all animals respond. Between these two extremes there is a graded response as shown in Figure 1. The shape of the curve is in the form of an elongated S and is normally called "S shaped". There are several interesting and important characteristics of this type of curve.

- Most of the curve is linear where the incidence of the response is directly related to the concentration. If the response is death there is no question that the compound is harmful in the dose range indicated.
- The concept of LD_{50} (the lethal dose for 50% of the animals) is a direct result of this curve. The LD_{50} is a statistical value and should be accompanied by some means of estimating the uncertainty in the value. This will be discussed in the next section dealing with normal distribution.
- The LD_{50} from the curve in Figure 1 is obtained by drawing a horizontal line from the 50% mortality point on the y axis to the point where it intersects the curve. From that point a vertical line is drawn to the x axis. The LD_{50} is the concentration represented by the point where the abscissa is intersected.
- The lethal dose for any predetermined percentage of the animals may be obtained in a similar manner. For example the LD_{90} may be estimated from the graph as shown in Figure 1.

Normal Distribution

The dose response curve in Figure 1 may be regarded as an accumulative normal distribution. In this situation the sensitivity of the animals responding

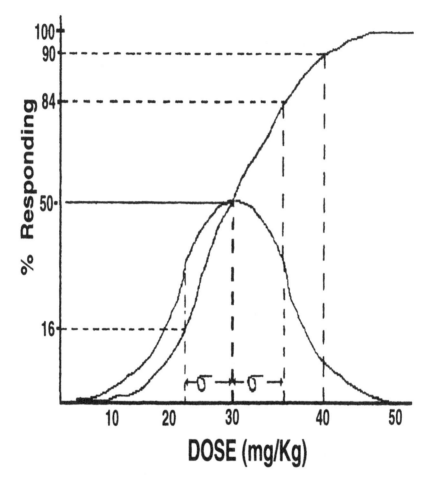

Figure 1.   A dose response curve for the data in Table 1. The bell-shaped curve demonstrates that the responses are normally distributed around the mean.

is normally distributed with respect to the drug dose, as shown by the bell-shaped curve in Figure 1. Therefore, at the low doses the sensitivity of the animals to the chemical is small. For example, at the very low dose none of the animals respond (Table 1). As the dose reaches the LD_{50}, the sensitivity approaches a maximum and then begins to decline, arriving at a point where the sensitivity is again small and all the animals respond.

In addition, the slope of the linear region of the dose response curve is related to the variance in the sensitivities. A steep slope represents a homogeneous sample of animals where the response to a change in dose is great. By contrast, a shallow slope represents a group of animals that are less homogeneous. The lack of homogeneity is reflected in the absence of sensitivity and response to a change in dose.

Table 2. Approximate LD_{50} for Selected Chemicals (Loomis, 1978)

Agent	Animal	LD_{50} (mg/kg)
Ethyl alcohol	Mouse	10,000
Ferrous sulfate	Rat	1,500
Morphine sulfate	Rat	900
DDT	Rat	100
Nicotine	Rat	1
Dioxin	Rat	0.045
Dioxin	Guinea pig	0.0006
Botulinus	Guinea pig	0.00001

This type of curve is similar to the normal distribution curves discussed earlier in Chapter 2. The same statistics apply. Thus, the median dose ±1 standard deviation includes 69% of the animals while the median dose ±1.96 standard deviations will include 95% of the results. The standard deviation may be estimated by drawing lines parallel to the x axis from the 50% ± 34.5% points on the y axis. At the point where they meet the dose response curve, lines are dropped perpendicularly to the x axis. These intersections on the x axis are 1 standard deviation from the mean or the LD_{50} (Figure 1). In the example shown in Figure 1 the LD_{50} of 30 mg/kg can be bracketed with ±11.7 (1.96 × the standard deviation of 6) to show the degree of confidence in the number. Unfortunately, the LD_{50} is usually shown as a single number with no indication of how sensitive the group of animals might be to the toxicant. Remember, the LD_{50} is not a precisely measured data point, but is subject to normal biological variability.

Scale of Toxicities

Potency of a chemical is related to the dose required to achieve the toxic effect under study. In the extreme, toxic response is expressed in terms of lethality. When describing the magnitude of the toxic effect the usual statement is that compound A is more (or less) potent than compound B. However, there are times when the evaluation of the toxicity of a single compound is required.[2] In other words, it would be desirable to make a statement like, "Compound A is very toxic" or "Compound B is harmless." As Table 2 indicates the absolute value of the LD_{50} may be in terms of a few micrograms per kilogram of body weight or as much as several grams per kilogram of body weight of

---

[2] There are many different measurements of toxicity. The LD_{50} is the most common and certainly the most frequently cited value. In this book toxicity and LD_{50} are sometimes used synonymously.

a particular compound. Chemicals that produce death in microgram[3] doses are considered extremely toxic (or poisonous). Other chemicals may be relatively harmless following doses in excess of several grams. Since a great range of doses are involved in the production of harm, categories of toxicity have been devised to allow comparison from one chemical to another. It must be realized that such classifications are very arbitrary. A typical one is shown below where the response is the dose killing 50% of the animals:

Extremely toxic	50 mg/kg or less
Moderately toxic	50–500 mg/kg
Slightly toxic	0.5–5 g/kg
Relatively harmless	5 g/kg or more

Margin of Safety

A result where the response is parallel to the y axis is impossible since that would mean a dose that produces no effect would simultaneously produce 100% response. Similarly, a result where the response parallels the x axis is also impossible since that would mean constant effect in the absence of chemical or in the presence of an infinitely large amount of chemical. Consequently, all compounds must have a slope lying somewhere between these two extremes. In fact, the slope becomes an index of the "margin of safety"[4] of a compound. The margin of safety is the range of doses involved in progressing from a noneffective dose to a lethal dose.

As the slope becomes more shallow, the margin of safety increases. Referring to Figure 2, compound D has a greater margin of safety than compound C. An equally correct statement is that the group of animals is less sensitive to D than to C.

Although the ultimate response resulting from a chemical agent is death, there are many other responses. In the case of drugs the response may be beneficial, e.g., controlling high blood pressure, relieving pain, etc. In developing new drugs the hope of the pharmacologist is to design a chemical that has beneficial effects at a much lower dose than the harmful effect. Thus, the margin of safety to the drug developer is the spread between the dose producing a lethal or undesirable effect and the dose producing the desired effect. The margin is referred to as the therapeutic index and is the ratio of the LD_{50} to the ED_{50} (effective dose for 50% of the animals). Figure 3 shows this ratio. As the lethal curve is shifted to the left, the therapeutic index becomes smaller. The goal of the drug designer is to maintain a high therapeutic index.

---

[3] 1 microgram (μg) = 0.000001 g or 0.001 mg. See Chapter 2 for a more detailed description of these prefixes.

[4] In the margin of safety, two chemicals are being compared using similar populations of animals. This is in contrast to the earlier discussion on variance and homogeneity where one chemical was being used on different groups of animals.

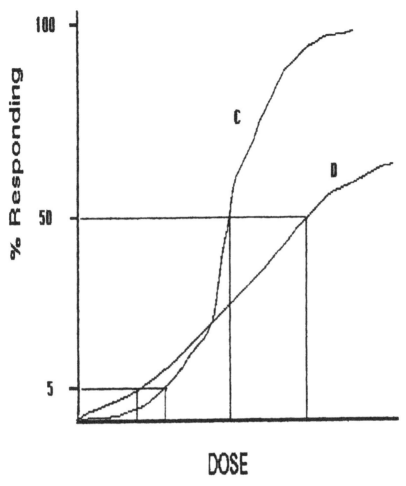

Figure 2.   **Hypothetical dose response for chemicals C and D administered to a uniform population of rats.**

BIOLOGICAL FACTORS IN TOXICOLOGY

Before a chemical can produce an effect on a biological organism, a reaction must occur. This means that the agent must contact some biological site and react. Therefore, the first task is to bring the chemical from the outside to the inside of the animal. Once inside, the chemical is transported to the key site where reaction takes place. Frequently these reactions are under the control of natural catalysts[5] which speed up or slow down the processes.

---

[5] These natural catalysts are called enzymes, which are proteins synthesized by the organism for this specific purpose.

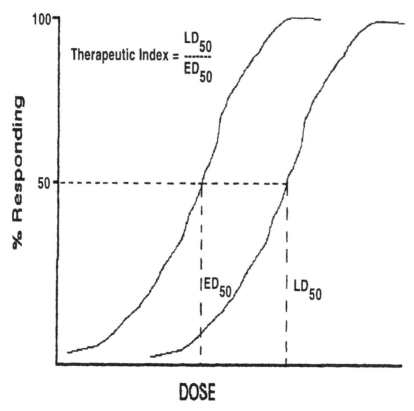

Figure 3.   Curves for a drug showing the therapeutic and lethal dose response effects.

Translocation of Chemicals

Animals are protected from the outside environment by specialized coverings. These protect the organism not only from temperature extremes, fluid loss, etc., but also block the free transfer of chemicals from the outside to the inside and the reverse. The coverings range in their degree of permeability from the skin (least permeable and the most protective of the layers) to the interior of the stomach and small intestine which are designed to be more permeable so food constituents once they are broken down can pass into the organism. A foreign chemical, as will be discussed later, has a much easier time of entering if taken orally. Once the chemical enters, a variety of translocation processes occur as shown schematically in Figure 4. These translocation processes expose the introduced chemical to several biological sites that might neutralize the molecule before the critical receptor is reached. Along the way the chemical may be biotransformed (or metabolized) to either a more active component or a fragment which can readily be eliminated.

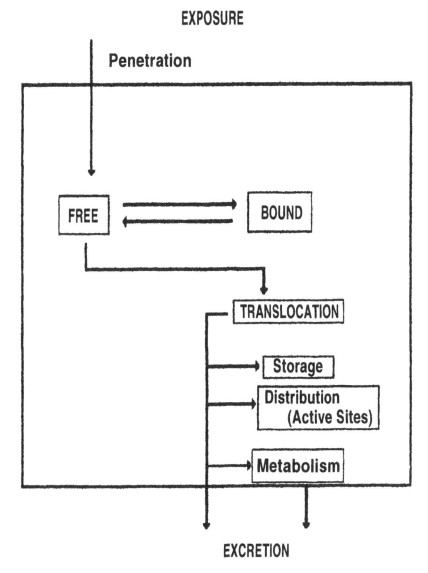

Figure 4.   Schematic illustrating the possible pathways a chemical can take once it has penetrated into the organism.

Storage of Chemicals

As Figure 4 demonstrates, a chemical can take many pathways inside the body. If there is a constant exposure, then eventually an equilibrium will be reached where the amount entering is equal to the amount being eliminated. The rate at which the body clears the chemical after the termination of the exposure depends on how fast the various sites will release the agent. For example, a chemical that is either tightly bound to a protein or has chemically reacted with a site will be cleared far more slowly than a material that is free. The drug Teridax[6] is so tightly bound to plasma protein that it takes 2.5 years to reduce the concentration by one half (Brodie, 1964).

From an environmental point of view, an important storage area is fat. Chlorinated hydrocarbons such as DDT and PCB[7] are very fat soluble. Once these materials enter an organism they are stored in the fat and slowly released back into the bloodstream. While stored in the fat, they are effectively removed from the system and will not cause any toxic response. A response will only be observed when the fat depots become saturated, resulting in blood levels high enough to cause an effect. Animals lacking fat depots will elicit the toxic response at much lower concentrations.

Tolerance

When repeated exposure to the same dose causes a decline in the response, the organism is said to have developed a tolerance to the chemical. Such tolerances would be graphically represented on the conventional dose response curve as a shift of the curve to the right. Tolerance commonly occurs with most habit-forming drugs. There are many examples that include cigarette smoking, alcohol consumption, and the use of narcotics such as heroin. Tolerance to chemicals has significance in toxicology for it represents a mechanism whereby certain organisms are protected against the harmful effects of chemicals.

Comparative Toxicology

This is a field of toxicology that is beginning to receive more and more attention. The increased awareness is due to the recognition that toxicity testing is done primarily in rodents and the information needs to be translated to humans. While not a great deal is known about the subject, the fact that rats are not humans must be remembered; as a consequence, testing uses many different animals. By covering a wide spectrum of species the hope is that the distribution will include organisms that behave similarly to people in their response to chemicals.

As an example of the wide variation of toxicity examine the comparative LD_{50} values for dioxin (Chapter 9, Table 2), where the lethal dose ranges from

---

[6] Teridax is a registered trademark of the Schering Drug Co.
[7] DDT: Dichloro diphenyl trichloromethane, an insecticide used extensively in the 1950s. PCB: Polychlorinated biphenyl, an industrial solvent that became widespread in the 1960s and 1970s.

1 μg/kg for guinea pigs to 5000 μg/kg for hamsters. Besides the wide variation in toxicity across species exhibited by this and other chemicals, the fact that humans are far from a homogeneous population must also be considered. For example, what is deemed safe for one person might be fatal to another individual. These are a few of the reasons why a dose that is deemed safe in an animal population is usually divided by 100 to establish a safe dose for humans. Hopefully, as more is learned about the differences, toxicologists can begin to quantify the effects in humans with greater precision.

CHEMICAL FACTORS IN TOXICOLOGY

There are two main areas in which chemical factors are important. The first is concerned with the transport across a membrane while the second deals with the reaction at a key site within the cell. These two topics will be discussed below.

Transport

Crossing the cell barrier depends on chemical properties such as lipid solubility and ionization. Current evidence strongly suggests that the nonionized, lipid-soluble form of an organic molecule is the predominant form capable of passing through the lipophilic membrane of the organism. Recall that many chemicals exist in both ionized and nonionized form.

$$HAc \longleftrightarrow H^+ + Ac^-$$

acetic acid	proton acetate ion
(nonionized)	(ionized)

The above reaction is reversible and the degree to which the reaction proceeds depends on the acidity or alkalinity of the solution. Accordingly, environmental conditions that favor the formation of the nonionized species will increase the rate at which molecules transfer across the various cell barriers.

Reaction

If the reaction between the chemical and the biological receptor causes a chemical change, the reaction is termed a biotransformation. These changes are quite common; consequently, in the determination of the toxicity of an agent, the effects of both the parent material and all fragments that might result from biotransformation reactions need to be assessed. Sometimes the transformed chemical is more toxic than the parent molecule. For example, the pesticide parathion is inert until the insect converts it to oxythion, a very toxic chemical.

Microsomal enzymes catalyze many of the biotransformation reactions. These are present in a variety of tissues, but are particularly abundant in the liver, making this a key organ for transforming different chemicals. The total quantity of microsomal enzymes can be increased in humans and other higher animals by prior exposure to a large variety of agents. For example, prior exposure to materials such as chloroform, DDT, barbiturates, etc. can cause an increase in the enzymes and therefore an increase in the ability of the system to alter the parent molecule. Obviously, in testing chemicals the previous exposure history of the animals in both the test group and the controls must be known.

ROUTE OF ENTRY

The route by which materials in the environment gain entry into the organism will depend to a large extent on the chemical and physical properties of the agent. Thus, volatile species will enter via the lungs while chemicals dissolved in water will undoubtedly enter orally. The four most common entry points will be discussed below.

Percutaneous Route

The simplest and most common exposure to foreign chemicals is by exposure through accidental or intentional contact with the skin. This is termed percutaneous absorption and is the transfer from the outer surface of the skin through the horny layer and into the blood circulatory system. Most often the skin is not a very efficient route for administering a chemical. This is illustrated by the observation that the comparative LD_{50}s for DDT in rats for the oral and dermal route are 118 and 2510 mg/kg, respectively (Loomis, 1978). A variety of factors such as ionization, molecular size, and water solubility are all involved in determining the ease with which a chemical can penetrate the dermal barrier.

Inhalation Route

For chemicals to reach the respiratory tract they must be either gaseous or sufficiently small so that they are not removed in the airway passages of the lung. The inhalation route is obviously the main route of entry for all airborne materials. While most of the atmospheric pollutants are no more than nuisance materials, others such as SO_2 and ozone can cause local toxicity.

Table 3.   Permissible Exposure Limit for a Representative Group of Chemicals

Material	μmol/mol	mg/m^3
CO_2	5000	9000
$CHCl_3$	25	120
Ethanol	1000	1900
Ozone	0.1	0.2
Sulfur dioxide	5	13

Many industrial workers are exposed to a variety of different volatile chemicals in the workplace in addition to the outside atmosphere. Because of the adverse effects of some exposures, standards establishing the concentration to which workers can be exposed during an 8-hour working day have been prepared. The National Institute for Occupational Safety and Health (NIOSH) has a continuing program for developing such standards. While NIOSH makes recommendations, it is the Occupational Safety and Health Administration (OSHA) that finally sets the standards that must be met. Each standard represents a consensus formed by a group of knowledgeable persons. The resulting document specifies the maximal allowable concentration in the workplace air for humans exposed in an 8-hour day and a 40-hour week. Industry has the obligation to establish conditions to comply with the standard. Presently, OSHA has set standards for several hundred chemicals ranging from inorganic materials such as beryllium (0.002 mg/m^3) to organic chemicals similar to toluene (375 mg/m^3).

Table 3 lists a few examples of the permissible exposure limit (PEL)[8] as the standards are commonly called. PELs are not intended for use in evaluating community air pollution. They are designed for the workplace environment. However, PELs undoubtedly will continue to be used to make decisions on atmospheric air concentrations in the nonworkplace environment.

Oral Route

The oral route is the third most common means by which chemicals enter the body (Loomis, 1978). This occurs when food contaminated with chemicals such as pesticides are eaten. The gastrointestinal (GI) tract, through which the food is taken, begins with the mouth and ends at the anus. While the chemicals are in this tube they are considered to be outside the body. Conse-

---

[8] In earlier years the group known as the American Conference of Government and Industrial Hygienists (ACGIH) developed consensus standards. These were referred to as TLVs (threshold limit values). OSHA has taken over this responsibility and issued standards that are now known as PELs; many of the former TLVs are now called PELs.

quently, most orally administered chemicals have an effect only after absorption has occurred from the GI tract into the body. While the chemical can be absorbed any place along the GI tract, the major site is the stomach. After absorption, translocation to the liver occurs; there the chemical is transformed (metabolized) either into harmless metabolites and excreted or into more toxic materials. Through these types of reaction the oral route can enhance the toxicity of a chemical compared to other routes of administration.

Parenteral Route

The final route of administration is not important from an environmental point of view. However, a brief discussion is included to complete the picture of how chemicals enter a biological organism. This route involves injecting chemicals directly into the body and bypassing the natural openings such as the mouth, nose, and skin. Pharmacologists employ this technique to introduce a known quantity of chemical to a specific site. Thus, the most rapid means of achieving a high concentration of chemical in a given location is to inject the chemical directly into the tissue by the parenteral route.

TESTING PROCEDURES

Deciding the effect of a chemical on a biological organism is a complicated procedure. To address this question a battery of tests are performed beginning with the simplest and cheapest and proceeding to the most costly and time consuming (Table 4). At the conclusion of all these tests the analyst is still faced with the most difficult and uncertain task of interpreting the data in terms of effects on humans.

The list in Table 4 includes the types of animal testing protocols being used to investigate the toxicological properties of a chemical. The table is arranged from the least costly to the most expensive. Usually the cost is in direct relation to the time involved in the test. The three main categories are the acute test, the subacute or prolonged test, and the chronic test. The acute involves the administration of one dose of the test chemical and waiting for a response, i.e., the acute lethal test that has been described in the earlier part of the chapter. In the subacute test the chemical may be administered daily for a period of up to 90 days. In contrast, the chronic test exposes the animal to a regular dose for most of the animal's lifetime. For example, a chronic test for a carcinogen involves the feeding of a low-level dose for the lifetime of the animal to see if such an exposure will cause the development of tumors. As one might suspect, the maintenance of the animals for this length of time adds a great deal to the cost. Most chemicals that are produced by industry

Table 4.   Testing Procedures

Early stage of development	Acute oral LD_{50}
	First screen
	Mutagen/carcinogenicity
	Subacute—90-day oral feeding
	Birth defects, teratology
Second level	Mutagen/carcinogenicity
	Metabolism and distribution of the chemical
	Residue evaluation
Late stage (chemical shows sign of commercial possibilities)	Long-term studies
	Carcinogenicity, teratogenicity
	Metabolism—humans
	Epidemiology—usually on existing products

are subjected to the early stage testing. Only those that are commercially important will have the full battery of testing, including chronic testing.

A few definitions will help explain some of the test procedures listed above.

Teratogenicity is the study of the effect of physical, chemical, and infectious agents on the developing embryo and fetus. This testing is very expensive, as is all the "icity" testing. The expense results from the fact that the tests are time-consuming and take a great deal of manpower to perform. Testing requires at least two species of animals with 20 pregnant animals per group. In addition, four dose levels are used.

Mutagenicity is concerned with the ability of a chemical to alter the genetic material of the cell. Some mutations can be helpful while others can be very harmful. The insidious part is that the effects will only appear in future generations.

Carcinogenicity is the induction of malignant neoplasms by the interaction of an outside chemical on a cell. The next chapter will address this topic in depth.

Epidemiology is the study of human populations to establish correlations between environmental exposure of chemicals and specific health effects. These types of studies are very difficult to conduct. The difficulty is in selecting two groups that are identical in all respects except for the factors under study.[9] However, when studies are performed under the guidance of a professional epidemiologist, the results can be very informative.

---

[9] The difficulty is well illustrated by the long time it took to establish a correlation between smoking and lung cancer.

TYPES OF TOXIC EFFECTS

A wide variety of effects are looked for during a toxicological investigation. A few that are important from an environmental exposure are

- *Allergic agents*, which cause itching, rashes, or sneezing.
- *Asphyxiants*, which cause displacement of oxygen and thus suffocation.
- *Irritants*, which cause pulmonary edema (fluid in the lungs) when inhaled at high concentrations and rashes when spilled onto the skin.
- *Necrotic agents*, which cause cell death.
- *Cancer, mutations, and deformed embryos*, which are the result of chronic exposure to low levels of carcinogens, mutagens, and teratogens, respectively.
- *Systemic poisons*, which can have an adverse effect on the whole body when taken internally.

Besides these general categories two other concepts are important, synergism and antagonism. The former describes the enhancing effect of one contaminant on another. A well-known example is the synergistic effect of asbestos and cigarette smoke. Either one will cause lung cancer but together the incidence of lung cancer is increased by two orders of magnitude. Antagonism describes studies in which the addition of a second compound neutralizes the effect of the toxicant. One example is the ability of atropine to neutralize the effects of cholinesterase-inhibiting agents such as the organophosphate insecticides.

4   CANCER AND
CARCINOGENIC CHEMICALS

INTRODUCTION

Among the effects of chemicals on biological systems (Chapter 3), one of the most feared is the initiation of cancer. The presence of such chemicals (known as carcinogens) in the environment has become synonymous with environmental contamination. Consequently, a few things need to be learned about the disease to place the contamination in proper perspective. This will be accomplished by discussing the subject under the following four categories:

1.   Statistical analysis
2.   Mechanism of cancer initiation
3.   Risk assessment
4.   Conclusion

STATISTICAL ANALYSIS

Cancer is an ancient disease, having afflicted our ancestors throughout history. For example, Egyptian medical records (3500 years old) describe diseases that would be characterized as cancer by today's criteria. However, deaths from cancer in those ancient times and even 100 years ago were rare. In 1850, about 1 in 200 deaths in the U.S. was caused by cancer. Today the rate is increasing, accounting for 1 in 5 deaths. Much of the modern increase in mortality from cancer is a result of a decrease in deaths from infectious diseases. A century ago, ailments such as tuberculosis, influenza, pneumonia, and diphtheria were all common and usually fatal. Since then, general acceptance of the germ theory led to a series of medical advances that began to reduce the frequency and severity of infectious diseases. These advances included filtering and chlorinating municipal water supplies, improving the

43

Table 1.   Causes of Death Expressed as a Percentage of Overall Death
Rate/100,000 Population

Year	Death rate	Percent				
		Cancer	Heart	Influenza	Gastritis	Auto
1900	1720	3.7	20	11.7	11.3	—
1910	1470	5.2	25.3	10.6	7.8	0.12
1920	1300	6.4	28.1	15.9	4.1	0.79
1930	1130	8.6	37.5	9.3	2.3	2.4
1940	1080	11.1	44.9	6.5	0.95	2.4
1950	960	14.5	53	3.2	0.53	2.4
1960	954	15.8	38.8	4.4	—	2.2
1970	945	17.2	38.3	1.8	—	2.8
1980	869	21.1	38.3	0.3	—	2.8

removal and treatment of sewage and other urban wastes, pasteurizing commercial dairy products, and most recently the development of highly effective antibiotics. Thus, as medical science learned how to control early childhood maladies, more people have lived through childhood and ended dying from cancer and other degenerative diseases such as cardiovascular disorders. Table 1 (Bureau of the Census, 1982) illustrates how the death rate has changed from various causes over the years.

How big is the cancer problem? As Figure 1 shows the historical life expectancy at birth has improved dramatically. A white infant born in 1900 would expect to live fewer than 50 years and most likely die of an infectious disease. By 1980, life expectancies at birth increased to 70+ years. With this increase has come a corresponding increase in the death rate from cancer as shown in Figure 2. This is the type of data that is used to prove the claim that the problem of cancer is increasing at an exponential rate.

While death rate from cancer is increasing further examination of the data is necessary. Table 2 shows that in the same period there has also been a marked increase in the number of people living to 65 and over. It is in this older population where the incidence of cancer is the greatest (Table 3). Figure 3 displays the exponential rise in death rate from cancer with increasing age.

A word of caution is necessary when examining statistical tables. Adjustment for age and other factors is very critical when surveying groups from different time periods, since living standards have changed dramatically over the last few decades. Thus, in comparing cancer death rates from previous years with the present, all the changes that have occurred in the intervening time period need to be recognized to make the groups similar in all aspects except the factor being studied. This is in line with the discussion on experimental design (Chapter 2) where attention was focused on control groups. For example, the data in Figure 2 are the crude death rates from cancer. When

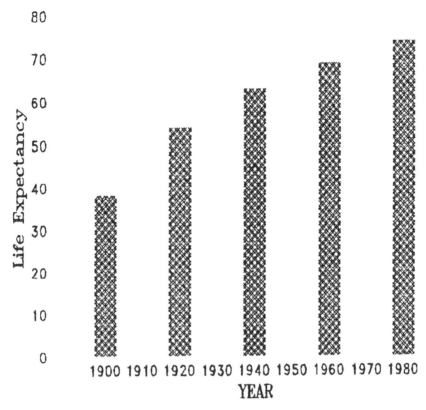

Figure 1.   Life expectancy of people born in the U.S. between 1900 and 1980.

the data for the years 1970 and 1980 are corrected for age, the rate is 130/100,000 in 1970 and 132/100,000 in 1980. This is a negligible increase compared to the uncorrected rates of 163/100,000 and 183/100,000 for the same two periods (Bureau of the Census, 1982). Improved ability to diagnose and report deaths from cancer also must be considered when comparing earlier years to the present time. A reasonable assumption is that in 1900 there were deaths from cancer that were not reported. For example, many people who contract cancer eventually die of pneumonia and this is the cause placed on the death certificate. Greater effort is being made today not only to improve diagnoses but also to record the actual causes of death on the certificate.

While progress has been made in the control of infectious diseases, little or no control has occurred in chronic disorders such as cancer and cardiovascular diseases. What has occurred, according to the data in Tables 1 and 3, is that a large fraction of the population is now living long enough to succumb to these chronic diseases. Life expectancy at age 65 and over has changed very little during the era of the greatest progress both in medicine and the standard

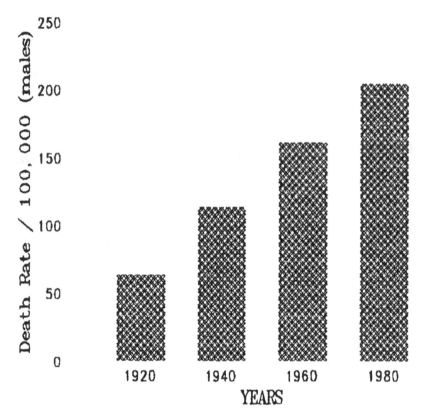

Figure 2.   Death rate from cancer in the U.S. (1920 to 1980).

Table 2.   Number of People in the U.S. Living in the 65+ Age Group

Year	65+ (millions)	Total population (millions)	65+ (% of total)
1900	3.1	76.1	4
1910	3.9	92.4	4.2
1920	4.9	106.5	4.6
1930	6.6	123.1	5.3
1940	9.0	132.1	6.8
1950	12.3	151.7	8
1960	16.4	179.3	9.1
1970	20	203.2	9.8
1980	25	225	11.0

Table 3.   Death Rate/100,000 Due to Cancer for Selected Age
Groups (1970)

Age rate	Death
25–44	26.2
45–54	189.6
55–64	520.7
65+	1364.3
Total all age groups	205.6

of living. All members of an aging population become debilitated and finally
die in very predictable and characteristic ways. It follows that the basic aging
processes in the population should be viewed as comprising a 100% fatal
disease. All the above factors have contributed to the increased attention
placed on cancer and possible causes of cancer. There are more people who
are at risk simply because as a nation we are living longer.

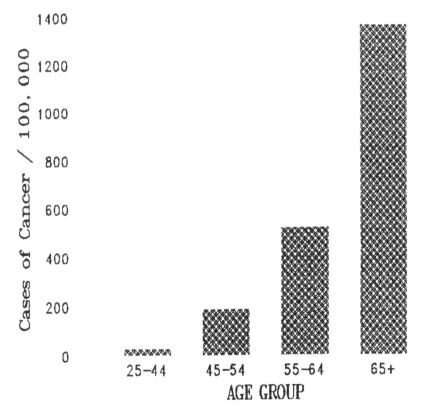

Figure 3.   Number of persons/100,000 in each age group who died of
cancer in the U.S. during 1979.

The above arguments are not meant to ignore the possible implications of being exposed to a chemical carcinogen in the environment. If the prevention of death from cancer can be accomplished by reducing exposure, then obviously the problem needs to be addressed. On the other hand, resources should not be spent needlessly if they will not significantly alleviate the problem.

MECHANISM OF CANCER FORMATION

Cancer, like all organic diseases, results from the malfunctioning of one of the many cells that make up the body. Cancers generally begin when a single cell is converted into a cancer cell by a process that is still not fully understood. Each descendant of the single cancer cell is also a cancer cell and all, in turn, produce more cancer cells. The malfunctions that occur are fundamentally different from the malfunctions in other diseases. In the latter case the interaction of the agent with the cell causes the death of the cell, e.g., polio viruses destroy nerve cells, bacteria that cause tuberculosis destroy lung cells, etc. Cancer cells by contrast are not injured nor are they killed. In fact, they are remarkably healthy. Two important characteristics identify cancer cells. First, they grow and divide with less restraint than normal cells and, second, they do not differentiate[1] normally and therefore do not perform their normal functions in the body. Growth and division can be understood through a discussion of turnover numbers. There are about 100 different cell types in the body that together add up to 300 trillion (300×10^{12}) cells. Some cells such as nerve cells live for many years. Others such as white blood cells have lifespans of only a few days. An intermediate type is the red blood cell, which lives about 120 days. The body contains about 25 trillion red blood cells. Thus, on the average, a healthy adult forms 25 trillion red blood cells every 120 days. This translates to about 2.5 million cell divisions every second. Biologists speak of this as turnover and say that red blood cells turn over at the rate of 2.5 million per second (Prescott and Flexer, 1982). In all cell types where turnover is present, the rate of cell reproduction is closely regulated. Disturbances in these controls result in disease. Cancer results when the disruption in controls affects reproduction and differentiation. In particular, cancer cells do not die on schedule. Eventually, the overgrowth interferes with the activities of normal cells resulting in the death of the organism. For example, certain white blood cells circulate for a few days and then die. Dying is as much a part of the process as growth. In certain forms of leukemia, the cells do not differentiate in a typical manner. One result is that the life cycle of a leukemia cell is prolonged. As they accumulate and crowd out the regular cells the body becomes less able to defend itself against infections by bacteria, virus, and other pathogenic agents. Understandably, the immediate cause of

---

[1] Differentiate: to become different or become specialized.

death associated with leukemia sometimes results from a bacterial or viral infection.[2]

What causes the conversion to a cancerous cell? Previous discussions established that cancer is transmitted from cell to cell through the process of cell division. The primary conversion of the first cell must be related to a hereditary defect or through a direct interaction on the cell by some foreign agent (such as radiation, chemical, etc.) or through a chance mutation that might occur at the time of cell division when the DNA strand is exposed. After the primary initiation,[3] the process is governed by the same genetic mechanisms that occur in all cells. The most obvious clue to the genetic basis of cancer is the observation that every descendant of a cancer cell is itself cancerous. In other words, once the original conversion has taken place, the genetic mechanisms involved with growth take over and faithfully reproduce exact copies of the original cancerous cell. Another important fact established beyond question is that radiation and certain chemicals can cause cancer. These same agents are also efficient inducers of permanent genetic changes, i.e., they cause mutations by interacting with the actual DNA in the gene. Representing the genetic material by DNA, the sequence of events leading to a cancer can be visualized as follows:

$$\text{DNA} \underset{k_2}{\overset{k_1}{\longleftrightarrow}} \text{DNA-x} \overset{k_3}{\rightarrow} \text{C} \tag{1}$$

where

          DNA   = deoxyribonucleic acid in the gene
    DNA-x = altered DNA through interaction with a chemical, radiation, or a spontaneous mutation
           C   = cancerous cell
        k_1 = rate constant for conversion of DNA to DNA-x
        k_2 = rate constant for repair of altered DNA
        k_3 = rate constant for conversion to cancerous cells

Cancer may be initiated (process represented by k_1) in a variety of ways. Three of the more important are

1.   By chemical interaction with DNA forming a new structure DNA-x, where x is a covalently bound fragment.
2.   Alteration of the normal DNA by radiation can give rise to an aberrant form DNA-x. (Note these two DNA-x's are different struc-

---

[2] It is for this reason that death certificates sometimes report the immediate cause of death as viral or pneumonia and fail to recognize the role of cancer.

[3] The scientific community at the present time favors a multi-step process in the initiation, as opposed to a one-step process such as a single mutation.

3. In the normal process of cell division, the DNA in the act of dividing may give rise to a random aberrant form. These new forms of DNA can form cancerous cells or give rise to other abnormal reactions. In fact many people describe the process of aging as a collection of mistakes caused by the formation of altered DNA.

The body has several defense mechanisms (process represented by k_2) which tend to counteract the formation of DNA-x and remove the altered molecule before the mistakes can do any damage to the host system.

1. An immunological mechanism where the system recognizes the abnormal molecules and neutralizes them before they can begin to multiply.
2. Repair mechanisms where enzymes convert the altered DNA back to a normal DNA.

When one of these repair mechanisms is out of control tumors may develop. The time between formation of C and the appearance of the visible tumor is called the latent period (Figure 4). This period may be shortened by any mechanism that causes a speedup in tumor development. The length of this latent period, which may be years, makes detection of cause and effect very difficult. By the time a tumor is visible there is no record of the initiating event.

Single exposure to a carcinogen may, in some individuals, be enough to cause the appearance of a tumor. For example, if an individual has a low immune defense mechanism, the appearance of DNA-x may cause the manufacture of more cells containing DNA-x which will cause the appearance of tumors.

With this understanding of the various mechanisms of cancer formation has come a scheme for classifying chemical carcinogens into two categories, strong and weak. Strong carcinogens are chemicals that react directly with the DNA, creating an aberrant genetic template. This is the reason they are sometimes called initiators rather than strong carcinogens. The replication of this template constitutes a mutation, and provides the potential for the development of a malignant tumor.

Compounds classified as weak carcinogens do not react chemically with DNA. Examples are chloroform, saccharin, vinylidine chloride, etc. These materials, because of their toxicity, promote tumors. For this reason the class of chemicals has been called "promoter." Once the cell is in a stressed condition, the frequency with which an aberrant DNA molecule may be formed is increased. The implication is that the hazard of cancer from short-term exposure to weak carcinogens is limited. Further limitation occurs if the exposure is sufficiently low to avoid cellular toxicity. An important characteristic of

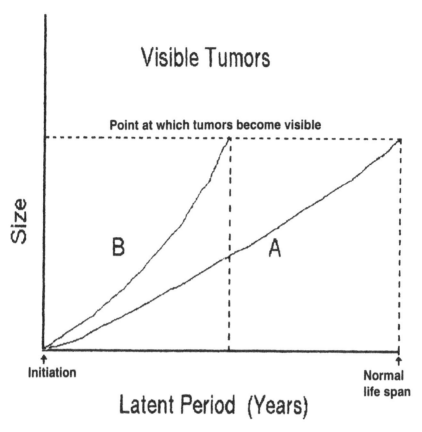

Figure 4.   A hypothetical plot showing the relation between initiation of cancer and the appearance of a visible tumor. Curve A: normal population; curve B: where the latent period has been shortened.

promoters is that the animal requires a long exposure at a dose approaching the maximum tolerated dose before the appearance of tumors is noted.

RISK ASSESSMENT

Classifying carcinogens as promoters and initiators has created two schemes for performing risk assessment. While these schemes have not been officially accepted by government regulators, more scientific investigators are beginning to accept them. Each will be described using a hypothetical database to illustrate the details.

Table 4. Hypothetical Cancer Data to Illustrate the Risk Assessment Models

	Control	Dose (µg/kg/day)			
		0.001	0.01	0.05	0.1
Tumors/50 animals	4	3	20	22	30
Response	0.08	0.06	0.40	0.44	0.6

The Data

A chemical has been found to generate cancer in rats when fed in the diet over the lifetime of the animal. The following results were obtained (Table 4).

It should be noted that the controls exhibited tumors (4 out of 50 animals)—very rarely is an experiment performed when no cancer is observed in the control population. The dose of 0.001 µg/kg/day is called the no observable effect level (NOEL).[4]

Promoter

On further study the determination was made that the dose necessary to achieve the appearance of tumors approached the maximum tolerated dose. In other words, any dose above 0.01 µg/kg/day would cause toxic symptoms other than cancer. From this and many other observations the conclusion was reached that the chemical was a promoter.[5] Accordingly, by maintaining the exposure well below the NOEL a safe level could be estimated. This was done by taking 1/100 of the NOEL (Table 4) or 1/100 of 0.001 µg/kg. The factor 1/100 is to convert rat data to human data with a safety factor in favor of man. Thus, 1×10^{-5} µg/kg/day is assigned to the RSD (risk-specific dose). For a 70-kg person this equates to 7×10^{-4} µg/day. The interpretation is that by keeping the daily intake at or below 1×10^{-5} µg/kg/day the individual will not contract cancer. However, cancer can still be contracted from other sources. From the statistical tables cited earlier (Table 1, Figure 2, Table 3, etc.) the chance is still 1 in 20 that an individual might die of cancer. Remember, death is a certainty (probability of one) and the probability cannot be eliminated simply by keeping the level of a suspect carcinogen below the RSD.

---

[4] Following the discussion in Chapter 2 a group of 50 animals is large enough to detect a carcinogen at the 16% level with a degree of confidence of 95%. However, it should be noted that, if the dose is decreased and it is desired to detect a difference at the 2% level, then a group of 1237 animals would be required.

[5] In practice a great deal of research is required to distinguish a promoter from an initiator.

Initiator

When cancer is caused by means of an initiator, the question of threshold is not as significant as with the promoter. The reason is that once DNA-x has been formed (Equation 1) the risk of cancer becomes a finite possibility. Consequently, there is no threshold below which cancer is not induced. As the dose is lowered, the probability becomes smaller but never zero. In other words, there is always a risk of developing cancer after exposure to an initiator. The significance of this risk at these low doses becomes a difficult problem to quantify. Since the rate can only be decided statistically, the difference between the incidence of cancer from people exposed to a small dose and people who have not been exposed is impossible to measure with any degree of precision. This is true because of the natural background incidence of cancer which shows up in most control groups. One way out of this dilemma is to use the measurable increase in cancer caused by a high dose to estimate how much increase might be expected at low doses. If a given dose causes a 20% increase, then 1/100 of a dose might be expected to cause 0.2% increase. Current controversy over setting standards revolves around the validity of this assumption.

The problem of quantifying risk through the extrapolation of experimental data has led to the development of several mathematical models. Since the biological mechanism is not known, these models become exercises in curve fitting. This is well illustrated by examining the data in Table 4 as shown in Figure 5. Since the point of zero response and zero dose (i.e., the origin in Figure 5) is assumed to be part of the data set, the problem is to find an equation that includes all three experimental points and passes through the origin. As can be seen from Figure 5 this involves a range of several orders of magnitude from the experimental data to the assumed zero point. Once such an equation is developed, the risk can be estimated for any level (Park and Snee, 1983). Currently, the EPA uses a risk of 1 in 1 million (1×10^{-6}) as acceptable. Certain states in deriving water quality levels accept a risk of 1 in 100,000 (for example, see Chapter 9, the case study on dioxin).

Most of these models (Park and Snee, 1983) reduce to the simple intuitive relation shown in Equation 2 for small doses.

$$\text{Risk(d)} = b \times d \tag{2}$$

where

$R(d)$ = the probability of acquiring cancer at dose d given that no cancer would have been observed in the absence of the dose; this is called the additional risk over background

b = slope of the risk(d) vs. d at low dose

Risk in this model is the additional lifetime risk of developing cancer from exposure to the defined dose over the lifetime of the animal.

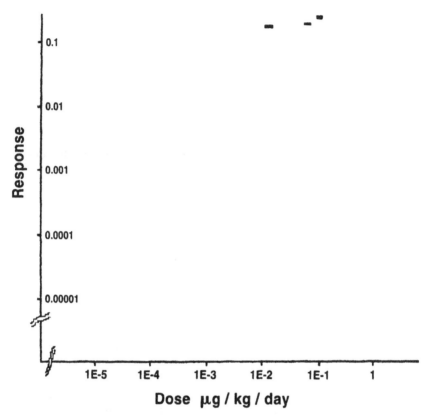

Figure 5.   Plot of the data shown in Table 4. The break in the x and y axis indicates that zero dose on this scale is never reached.

Computer programs have been developed to analyze the raw biological data with sophisticated statistics. This analysis depends on extrapolating from a region of known response 6/10 (taken from the data in Table 4) to a risk of 1/1,000,000—an extrapolation of four orders of magnitude! While there is something comforting about feeding numbers into a computer, never forget that extrapolating data into regions where there are no data has no biological validity.[6]

The calculation that will be illustrated is an approximation of what the computer models will do. The procedure is shown below.

1.   The first step is to correct the response data for the tumors observed in the control group by using Abbott's formula. Thus, at the high dose the new response becomes:

---

[6] The reason that such extrapolations are not valid is that the investigator has no basis for deciding what type of reactions occur in the unknown region.

$$R = (R - \text{control})/(1 - \text{control}) \times 100$$
$$= (0.6 - 0.08)/(1 - 0.08) \times 100$$
$$= 56.5\% \text{ tumor response at } 0.1 \ \mu g/kg/day$$

Since this is more than 50% tumor development the next lower dose is used. Correcting the response at the dose of 0.05 μg/kg/day by means of Abbotts's formula yields 39%. Using a response of 50% or greater is an arbitrary cutoff. As seen from Figure 5 if the slope was estimated from the 56.5% response it would yield a less conservative estimate for the risk. Consequently, a dose is chosen where the response is less than 50%. In this case it becomes the next dose, i.e., 0.05 μg/kg/day.

2.   The slope = risk/dose = 0.39/0.05
$$= 7.8(\mu g/kg/day)^{-1}$$

This is considered the cancer potential factor for rats. A correction is now made to convert rat data to human data. This is based on a surface area conversion that corresponds roughly to the ratio of the cube root of the weights. For a 70-kg person and a 0.4-kg rat the factor is

$$(70/0.4)^{1/3} = 5.8$$

In other words, humans are more sensitive than rats by a factor of 5.8.[7] The cancer potential for humans, sometimes designated as Q*, becomes

$$Q* = 5.8 \times 7.8 = 45.2(\mu g/kg/day)^{-1}$$

3.   For a risk of 1 in a million
$$\text{Dose} = \text{risk}/Q*$$
$$= 1 \times 10^{-6}/45.2$$
$$\text{RSD} = 2.2 \times 10^{-8} \ \mu g/kg/day$$

The literal interpretation of the calculated RSD is as follows. For a population of 1 million exposed to the RSD for 70 years, the upper limit on the increase in number of cases is 1 in 70 years. If the background incidence for the population is 25%, then 250,000 cancer deaths will be expected. With the additional risk it will be 250,001.

These types of numbers should only be used for ranking priorities and comparing risks from other exposures. Unfortunately, they are being used to estimate actual risks because it is the only number available. Again the use

---

[7] The cube root of the ratio of weights is an empirical relation based on many observations.

of the heuristic of availability (Chapter 1) is illustrated. Decisions are made based on the available data no matter how good or bad those data may be.

CONCLUSION

Currently there is no "magic bullet" for the cure of cancer. Accordingly, society must learn how to live with the presence of this disease without being unduly afraid. The purpose of this chapter is to place the subject in perspective and to treat it in an objective manner. By examining the national statistics, the discovery was made that cancer belongs to a class of diseases that largely affects our older population. Earlier in our history, before 1900, the early onset of childhood diseases such as pneumonia, influenza, etc. masked the rate of incidence. Now that these ailments are under control, more people are living long enough to succumb to chronic disorders, such as cancer and heart disease.

Without the "magic bullet," the control strategy for cancer has focused on decreasing exposure to known carcinogens. Before investing time and resources into decreasing the exposure, the extent of the risk needs to be understood. This has led to the technique known as risk assessment. In other words, what is the magnitude of the risk from the exposure to the chemical under discussion? An example of how such an assessment is performed was designed to illustrate the technique, and to provide a better understanding of the numbers and how they are used in estimating the risk. By applying this technique to actual exposures the reader is allowed to make his or her own decision. The case study on dioxin (see Chapter 9) will illustrate the method using real exposure and toxicity data.

5   THE ATMOSPHERE

INTRODUCTION

A few of the effects that exposure to chemicals may have on humans have been presented. The next step is to characterize the major environmental compartments containing the released chemicals. This chapter will focus on the atmosphere. Chapter 6 will deal with water.

The impact of chemicals on the atmospheric compartment may be divided into two main categories, macro and micro. As the name implies, macro effects occur when the introduction of chemicals can alter patterns on a large scale. Global warming and the destruction of the stratospheric ozone are two examples of such phenomena. To understand these effects, a study of the physical makeup of the atmosphere must be made. The section below will address this objective through a discussion of the influence that introduced chemicals have on the properties of this important compartment.

The second category, micro, is the more traditional. In this situation, local air parcels concentrate the introduced chemicals. When a concentration exceeds the level determined to be hazardous, the problem must be alleviated. This is performed by lowering the amount of chemical entering the air parcel or by breaking up the parcel so that the material becomes dissipated.

MACRO SCALE

Viewing the atmosphere from the surface of the earth, one can imagine being at the bottom of a limitless sea of air. However, the truth of the matter is the atmospheric pressure (hence, the concentration of the gases) falls off quickly with increasing altitude from 101.3 kPa (1 atm) at sea level to less than 0.13 kPa (0.0013 atm) at 50 km in altitude; 90% of the mass is below 15 km. This thin skin, similar in scale to the peel on an orange, is very critical. Without it, life cannot exist. Consequently, the layer must be protected.

Evolution of the Atmosphere

There is good evidence that the present atmosphere arose from conditions that were far different from what exists today. The first confirmation comes from astronomy.

Astronomy

According to prevailing theory, the solar system (including earth) was formed from the dust and gases of the universe. Therefore, the original earth's atmosphere would have contained the same volatile elements known to be abundant in the universe. Of the ten most common elements, three are noble gases,[1] helium, neon, and argon. Accordingly, these must have been present in the early atmosphere. Today, except for argon, they occur only in minor amounts. Present-day argon is thought to have originated from the radioactive decay of potassium-40 in the earth's crust, as opposed to being a holdover from the original mix that coalesced to form the planet. Since helium is the lightest of the three, it could have escaped the gravitational effect of the earth. If the same explanation is used for neon, then the temperature of the earth must have been hotter in order for the neon to have attained escape velocity.[2] Consequently, the atmosphere, if there was one, must have gone through some profound changes from a primitive system containing the noble gases to one that is now dominated by nitrogen and oxygen.

Sedimentary Deposits

The original atmosphere did not contain oxygen. This is supported by examining sedimentary deposits in South Africa which have been dated at 2 billion years. The deposits contained pyrite (FeS_2), a reduced form of iron that would have been oxidized to $Fe_2(SO_4)_3$ in the presence of oxygen. Since the deposits were in the reduced form the conclusion emerges that the original atmosphere lacked oxygen.

Photosynthesis

One important event in evolution was the appearance of oxygen about 2 billion years ago through the process of photosynthesis. By developing this

---

[1] This is the name given to the group 8A elements in the periodic table. These elements are characterized as being gases and also chemically unreactive.

[2] Escape velocity is the speed required (11 km/s) for any system, from molecules to spaceships, to overcome the gravitational pull of the earth.

mechanism, organisms were able to utilize water and carbon dioxide (the latter was present in much greater concentration than is found today), and through the action of sunlight synthesized sugar and molecular oxygen. The importance of this set of reactions is reflected in the energy that can be derived from sunlight and stored in the form of complex molecules such as sugar. Simultaneously and of equal importance was that the process produced oxygen, a necessary component of present-day life forms.

$$6CO_2 + 6H_2O \xrightarrow[\text{photosynthesis}]{\text{sunlight}} C_6H_{12}O_6 + 6O_2 \tag{1}$$

The energy stored in the sugar was recovered by breaking the sugar into smaller pieces:

$$\underset{\text{sugar}}{C_6H_{12}O_6} \rightarrow \text{lactic acid} + \text{energy} \tag{2}$$

Eventually, as the concentration of oxygen increased, more efficient organisms evolved which were capable of completely oxidizing the sugar by a process known as respiration. The overall equation is

$$C_6H_{12}O_6 + 6O_2 \xrightarrow[\text{respiration}]{} 6CO_2 + 6H_2O + \text{energy} \tag{3}$$

Normally, carbon is cycled between carbon dioxide (oxidized form of carbon) and organic carbon (reduced form of carbon). The reactions are under the control of photosynthesis (Equation 1) and respiration (Equation 3). The removal of organic carbon from the system by being buried in sediment[3] caused a net increase of oxygen in the atmosphere. Through this mechanism, the oxygen concentration was raised to the present level.

Present-Day Atmosphere

Homosphere

The atmosphere may be divided into two main areas, the homosphere and the heterosphere. In the first region the gases are uniformly mixed resulting in a constant composition. Table 1 indicates the main elements that are present (Ebbing, 1987). As indicated earlier, the density decreases rapidly with altitude.

---

[3] It was the trapped carbon that over the eons was transformed into the coal and oil. These deposits are presently being used as energy sources.

Table 1.   Composition of Gases in the Homosphere

Component	% by Volume	μmol/mol
Nitrogen (N_2)	78.08	780,800
Oxygen (O_2)	20.95	209,500
Argon (Ar)	0.934	9,340
Carbon dioxide (CO_2)	0.033	330
Trace Gases		
Chlorofluorocarbons		0.0002
Ozone (stratosphere)		0.03

From the data in Table 2 (Ebbing, 1987) not only density but also temperature and pressure are functions of height. The total mass of the atmosphere is approximately 5.2×10^{21} g (Weast, 1986).[4] Appendix II illustrates how a simple calculation yields this mass.

The Heterosphere

The heterosphere, which ranges from 90 to 10,000 km, is characterized by a separation of the gases based on mass. In order of decreasing mass the layers are made up of nitrogen (closest to the earth's surface) followed by oxygen, helium, and hydrogen. Hydrogen atoms that are still under the gravitational pull of the earth have been found at 30,000 km (18,000 miles). While this can be considered the outer boundary of the atmosphere, remember that the concentration is very dilute in these high altitudes, i.e., the density at the base of the heterosphere (90 km) is one millionth of the density at sea level.

Table 2.   Temperature, Pressure, and Density of the Atmosphere as a Function of Height Above the Earth's Surface

Height (km)	Temperature (°C)	Pressure (kPa)	Density (g/m³)
20	−51	5.86	92
10	−45.5	27.27	419
8	−29.7	36.48	524
6	−15.1	47.91	649
4	−3.0	62.06	803
0	15.7	101.3	1220

---

[4] The troposphere is defined as the lower atmosphere (about 10 km in height) where most of the wind circulation occurs. It is also defined by the height where the sharp temperature inversion develops as shown in Figure 1.

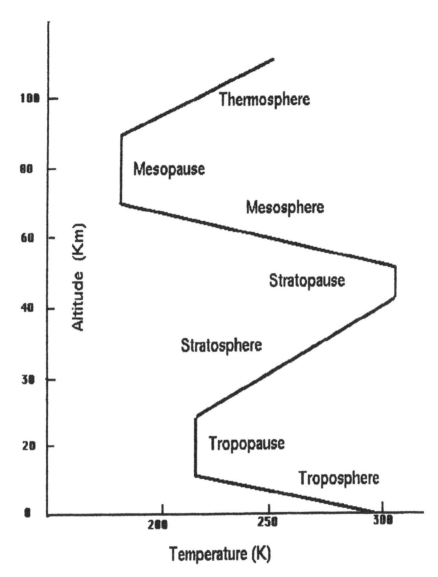

Figure 1.   Division of the atmosphere based on temperature.

Temperature Gradient

Another division can be accomplished based on the temperature profile shown in Figure 1. The changes in temperature with altitude can be explained in the following manner. As the sun's radiation enters the top of the thermosphere (120 km) the oxygen molecule absorbs ultraviolet radiation (wavelengths below 200 nm), breaking the bonds as shown in Equation 4.

$$O_2 + hv \rightarrow 2O + heat \qquad (4)$$

This is an endothermic reaction in which the liberated heat is added to the system, causing both an increase in temperature and in the concentration of oxygen atoms. As the rays from the sun pass through the thermosphere, ultraviolet radiation is dissipated. This reduces the impact of Equation 4, and the temperature of the region begins to fall. The lowering of the temperature continues until an equilibrium is reached at the mesopause. Dropping through the mesosphere, the temperature begins to rise again because of the increased pressure on the gaseous molecules. At the stratopause another temperature equilibrium is reached.

The stratosphere, like the thermosphere, is a region of chemical reactivity. The concentration of O_2 and O are sufficiently great to react in the presence of a third body (M in Equation 5) to form ozone.

$$O_2 + O + M \rightarrow O_3 + M^* \qquad (5)$$

M is a molecule such as nitrogen which takes away the excess energy as an excited molecule. Unless the energy formed in Reaction 5 is dissipated, the ozone will immediately dissociate back to O_2 and O. The reason ozone is not synthesized in the thermosphere is that the concentrations of the reactants are so dilute that the probability of a three-body (O_2, O, and M) collision is very remote.[5] Below the stratosphere the concentration of oxygen atoms (not to be confused with oxygen molecules) becomes negligible; hence, ozone is found in a very narrow band in the stratosphere (30 to 35 km).

The formation and destruction of ozone is shown in Reaction 6.

$$O_2 + O \rightarrow O_3 + infrared$$
$$O_3 + ultraviolet \rightarrow O_2 + O \qquad (6)$$
$$ultraviolet \rightarrow infrared$$

In this series of reactions, ozone is a catalyst for the conversion of ultraviolet radiation (UV) to infrared (IR) or heat, causing the observed temperature changes. By the time the radiation has reached the tropopause, most of the UV has been absorbed, leaving the troposphere as an unreactive region. This is of vital importance since the high-energy UV has the ability to destroy DNA, the genetic code of life.

Given that ozone in the stratosphere absorbs all of the UV between 200 and 300 nm, it is remarkable how little ozone is actually present. The amount is normally reported in terms of thickness with which the ozone would blanket the earth if it were present as a layer of pure gas under conditions of standard

---

[5] However, see the section dealing with the formation of smog. Here oxygen radicals are present in sufficiently high concentrations to react with oxygen and form ozone.

temperature and pressure (0°C and 1 atm). The stratospheric ozone layer in these units varies with latitude and season (Figure 1) but averages about 3 mm in depth.

Continuing the discussion of the temperature changes, Figure 1 indicates that the tropopause is an area of cold air relative to the air coming from below. In the troposphere, the air is warmer at the surface due to radiation. The warm air becomes less dense and begins to rise and cool. Upon meeting the relatively warm air in the stratosphere, an inversion layer is formed. This creates a large box that is about 10 km high (the approximate height of the stratopause or inversion layer) covering the surface of the earth. Within this box, turbulence causes a thorough mixing of all the components. The lid on the box is not perfectly sealed, and over a period of time there is a slow transport of gaseous molecules from the troposphere into the stratosphere.

In addition to the lid formed by the inversion layer at 10 km, the troposphere can be further divided into two compartments north and south of the equator. Thus, two tropospheric compartments are created by the northeast and southeast trade winds moving across the oceans and picking up moisture. Heavy rain results when the winds meet at the equator. The heat released by the condensation is the source of most of the energy required to maintain the global wind patterns. This region, known as the intertropical convergence zone (ITCZ), is synonymous with a huge furnace whose principal fuel is water. The rising column of warm air in the ITCZ serves as a slight barrier to gas molecules passing back and forth between the northern and southern hemisphere. On a macro scale, three major compartments can be visualized separated by barriers of varying degrees of porosity (Figure 2). The rate at which molecules move back and forth across the ITCZ is characterized by a residence time of about 1 year. This implies that uniform mixing between the northern and southern hemisphere takes about a year. The time delay in mixing is the reason that concentration differences are observed for certain gases north and south of the equator. For example, monitoring data on methylchloroform[6] for the years 1972 to 1977 indicated a higher concentration in the northern hemisphere (Lovelock, 1977). This is explained by the observation that most of the emissions occur in the highly populated, industrialized north. The slight barrier to mixing at the equator creates the disparity in the concentrations between the north and south. Once emissions are terminated, the concentrations would become equal in a year. The porosity of the inversion layer at the tropopause is much less, i.e., the layer is more impervious to the transport of molecules than is the ITCZ zone. On the average about 50 years is required for the tropospheric concentration to be reduced by half. The importance of this observation will become obvious in Chapter 9 when the case study dealing with ozone is presented. Briefly, the chlorofluorocarbons (CFCs), used both as propellants in a number of different packaging containers and as refrigerants,

---

[6] Methylchloroform (CH_3CCl_3)—a volatile halocarbon used as an industrial solvent.

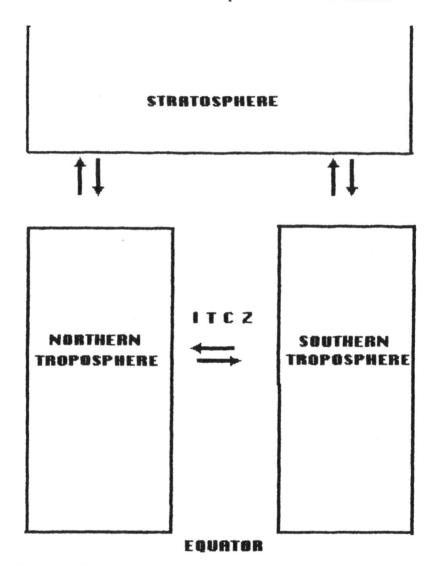

Figure 2.   **Illustrating the three compartments formed by the ITCZ at the equator and the inversion layer between the stratosphere and the troposphere.**

were allowed to escape into the atmosphere. Due to the barrier between the troposphere and the stratosphere, many years elapsed before these chemicals were transported into the stratosphere in sufficient concentration to disrupt a few key reactions involved in the synthesis and destruction of ozone.[7]

---

[7] These reactions are described in the ozone case study (Chapter 9).

Table 3.   **Active and Storage Deposits of Carbon**

	Capacity (10^{18} g)
Active	
As CO_2 or CO_3 in solution	
Atmosphere	0.69
Ocean—surface	0.52
Ocean—deep	34.81
As organic carbon	
Land organisms	0.45
Land—decaying	0.69
Marine organisms	0.01
Marine—decaying	3.01
Total	40.18
Storage	
As carbonates	20,410
As organic	
Fossil fuel	10
Total	20,420

Cycles

There are many natural cycles by which matter is circulated through the biosphere. Only two, the carbon and nitrogen cycles, will be described in this section. These are of great importance because human activities are beginning to influence the direction they take. Cycles need to be understood in order to minimize the impact and adapt our lifestyle to accommodate the changing conditions.

Carbon Cycle

The movement of carbon through the biosphere is critical since all of life is composed of carbon compounds in one form or another. Table 3 lists the active and storage pools (Bolin, 1970). Carbon is available for life in two main forms, carbon dioxide and organic carbon. The former is a minor constituent of air (Table 1) representing 1/30 of 1% of the total atmospheric volume. A much larger reservoir of CO_2 is dissolved in water (97% of the total CO_2). Most of the marine carbon is in the form of carbonate, but through Reaction 7, which is reversible, carbon dioxide is always available.

$$2H^+ + CO_3 \rightarrow CO_2 + H_2O \tag{7}$$

Carbon is also available in reduced forms such as carbohydrates and proteins.

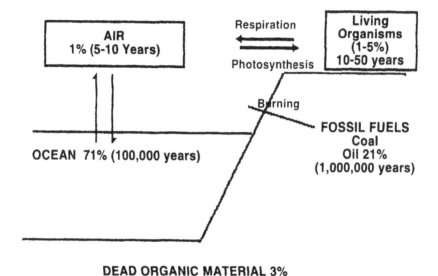

DEAD ORGANIC MATERIAL 3%

Figure 3.   Percentages of carbon in the various compartments (excluding sedimentary deposits). The numbers in parentheses are residence times.

However, by far the greatest amount of carbon is in storage as carbonate sediments, and is largely unavailable, as it takes millions of years to recycle.

Figure 3 illustrates the dynamics of the carbon cycle. In this and subsequent discussions the units are expressed as Eg (10^{18} grams or a trillion metric tons). The interchange between air and ocean is fairly rapid (0.11 Eg/year) and sufficient to replace the carbon dioxide in surface water every 6 to 8 years. The carbon in fossil fuels returns CO_2 through combustion. On the average, about 0.01 Eg of carbon per year are added to the atmosphere by burning. This has produced a significant increase in the amount of CO_2 in the atmosphere (Figure 4). The full implication of the climatic effects of this rising level of gas is still being debated as the following section illustrates.

Greenhouse Effect

There are many instances of how chemicals have an impact on the macro scale. One example dealing with the destruction of the stratospheric ozone layer will be discussed in detail in Chapter 9. Another example is the introduction of certain volatile chemicals into the lower atmosphere (troposphere) which block the escape of infrared radiation (heat) back into outer space. The burning of fossil fuels illustrates this situation. As demonstrated in Figure 4, the level of CO_2 is increasing in the troposphere. The problem associated with such gases

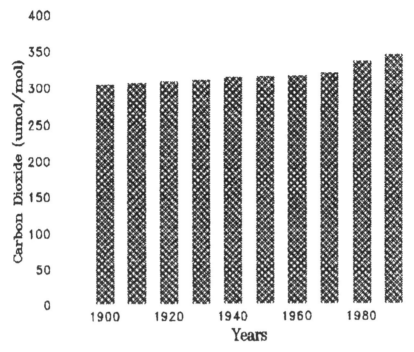

Figure 4.   Graph showing the increasing levels of CO_2 in the atmosphere over the years.

as CO_2, water vapor, CFCs, and methane is that they are nearly transparent to the visible and near-infrared wavelength in sunlight. However, they absorb and re-emit downward a large fraction of the longer infrared wavelength emitted by earth. The result is that the heat created by the reflection of sunlight from the earth's surface can't escape, and the troposphere warms up. A similar phenomenon occurs in a greenhouse; hence, the label "greenhouse" has been assigned to both the gases and the effect. Table 4 is a listing of the greenhouse

Table 4.   The Gases in the Atmosphere Contributing to the Greenhouse Effect (nmol/mol, ppb)

Gas	Pre-1900	1985
CO_2	275,000	345,000
CH_4	7,000	1,700
N_2	285	304
CFC-11	0	0.22
CFC-12	0	0.38
Methylchloroform	0	0.13
Cl_4	0	0.12

Note: The last four are manmade gases.

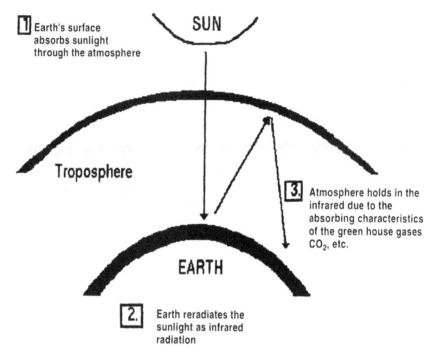

Figure 5.   Schematic showing the greenhouse effect in the atmosphere.

gases and the concentrations that are present in today's lower atmosphere (troposphere). While the CFCs are present in only trace amounts, they are 10,000 times more efficient than CO_2 at trapping heat (Revkin, 1988).

A schematic illustrating the radiation is given in Figure 5. Many reviews of this topic have appeared (Hileman, 1989; Brand, 1988; Lemonick, 1989). The problem has received so much attention that the EPA is now urging the U.S. and other nations to limit their emission of the greenhouse gases (Zurer, 1989). The magnitude of the problem may be visualized from the carbon balance shown in Table 5. While the respiration and photosynthesis are in balance, it is the increasing amount of CO_2 added by the burning of fossil fuels and tropical forests that is causing the most concern. The case study

Table 5.[9]   Data Showing How Carbon Dioxide Is Added and Removed from the Atmosphere (billion metric tons/year)

Added	Removed
Respiration 140	Photosynthesis 140
Burning of fossil 5	Absorption by oceans 2.5
Deforestation 1	Unknown 3.5

Note: If the trees of the Amazon were completely destroyed they would contribute 75 billion metric tons of carbon to the environment (Linden, 1989).

dealing with stratospheric ozone (Chapter 9) has a greater discussion of this worldwide problem and the actions that various governments are taking to alleviate the situation.[8] Present thinking is that atmospheric CO_2 is increasing by 3.4 billion metric tons per year while human activities add 7 billion metric tons per year. The issue of where the remaining 3.6 billion metric tons is located is still unanswered (Zimmer, 1993).

Nitrogen

Even though the atmosphere is mostly nitrogen the growth of many organisms is limited by a shortage of nitrogen. This shortage occurs because very few animals or plants can utilize atmospheric nitrogen directly. It must be converted to either nitrates (NO_3^-), nitrites (NO_2^-), or ammonia ions (NH_4^+). The nitrogen cycle provides a bridge between the large atmospheric reservoir and the forms that can be utilized by the organisms in the biosphere. Figure 6 shows a simplified cycle. One of the major concerns is that the fixation of nitrogen (converting N_2 to usable forms) is now exceeding the reverse process which is called denitrification. This causes soluble nitrogen to accumulate. The excess in forms of nitrate, nitrite, and ammonia salts is carried into the rivers, lakes, and oceans by means of runoff. As will be discussed in the next chapter, the increase in nitrogen speeds up the aging process whereby lakes are deprived of enough oxygen to support a desirable population of fish. This is a problem that will be accentuated in future years since the manufacture of fertilizer by fixing atmospheric nitrogen is doubling about every 6 to 7 years. Increased denitrification might ease the problem, if there was a greater understanding of the process and how it might influence other parts of the cycle.

MICRO SCALE

Micro-scale atmospheric effects are created by adding volatile chemicals into local air parcels until the critical concentration is exceeded and toxicity becomes evident. What are these local air parcels? Obviously, the most common is a room that is poorly ventilated. Such conditions can exist in the workplace, and many guidelines have been developed to help maintain a safe environment. For a more detailed explanation of these guidelines, see Chapter 3.

The workplace is the easiest area to monitor and control, since it is under the jurisdiction of both the owner and government regulators. A more difficult

---

[8] The two issues of stratospheric ozone and greenhouse effect are linked since the same chlorinated materials shown in Table 4 cause both effects.

[9] Rankin, R. A., in the *Detroit Free Press*, Oct. 30, 1989. More recent data (Zimmer, 1993) indicate that the total being added by man's activities is 7 billion metric tons.

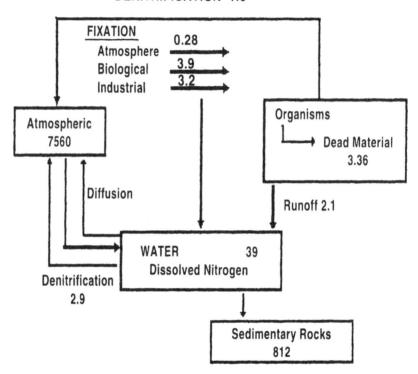

Figure 6.   Simplified nitrogen cycle. The amounts in the boxes are in exagrams (10^{18} grams). The rates of movement, associated with the arrows, are in 10^{12} moles/year.

air parcel to monitor is the home. The difficulty stems from the large number and the lack of control due to the adage that "one's home is one's castle" and no one should interfere. The case study dealing with radon (Chapter 9) describes a situation where indoor air pollution has become a troublesome situation.

A natural outdoor air parcel is formed when an inversion occurs. Normally the temperature gradient in the troposphere is from warm to cold until the tropopause is reached (Figure 1). During a temperature inversion, the colder air is at the ground and the warm air is above. Such a situation develops when a cold front (dense air) moves in under the warm air (less dense). With the temperature inversion, the upward current of warm air is blocked and any pollutants that are present stay in the air parcel near the ground. These effects are intensified by local topography such as that in Los Angeles. Here the cold air from the ocean comes in at night under the warm air created during the day. The inversion layer is formed and is prevented from moving on by the mountain range to the east of the city. Figure 7 illustrates this phenomenon and the temperature gradients that exist. This creates a situation in which tons

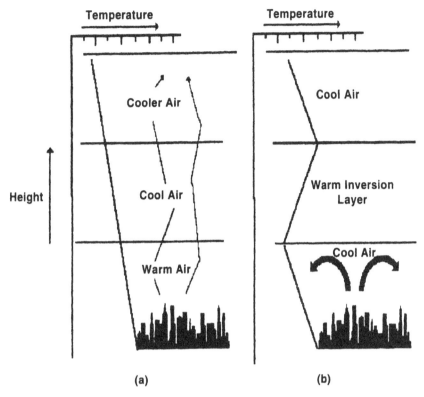

Figure 7.   **Temperature inversion. (a) Warm air rises, dispersing pollutants into the upper atmosphere. (b) Inversion: a warm layer
overlies cooler air and prevents pollutants from rising through
the warm air.**

of pollutants from automobiles, industry, etc. become concentrated in the local
air parcel and are baked by the sun's radiation, creating the infamous "smog."
Part of the reactions are due to the interaction of nitrogen dioxide emissions
from automobiles with sunlight as shown in Reaction 8.[10]

$$NO_2 \xrightarrow{\text{Sunlight}} O + NO \qquad (8)$$

$$O_2 + O \rightarrow O_3$$

The local buildup of ozone that occurs in urban areas such as Los Angeles,
Denver, Mexico City, New York, etc. during inversion has created a situation

---

[10] An expanded discussion of these reactions is given in *Chemicals in the Environment*
(Neely, 1980).

requiring stringent controls to be placed on auto emissions and many industries. Gradually the number of "alert days"[11] in the urban area is being reduced. At some point, a decision will have to be made regarding the expense involved in further reductions as opposed to the increased benefits in terms of improved health.

CONCLUSION

The purpose of this chapter has been to provide an overview of the atmosphere, and an appreciation of how introduced chemicals can influence the properties, both globally and locally. By making a few assumptions, the volume of air can be estimated. Once these volumes are known, reasonable approximations of the resultant concentration can be made. For example, knowing the area of the urban area and the height of the inversion layer, a volume can be calculated. Adding a mass of chemical to this volume generates a concentration profile. Matching the concentrations with toxicity data will then permit a preliminary assessment of the risk to be made.

Finally, the discussion of macro effects has provided a frame of reference for assessing the impact of human activities on atmospheric chemistry and how these disturbances can influence global patterns. The case study dealing with ozone in Chapter 9 describes one of these issues in greater detail.

---

[11] Alert days are the number of days when a city has exceeded the national ozone standard set by the EPA.

6   WATER

INTRODUCTION

The second major compartment that becomes a receptacle for synthetic chemicals is water. Unlike the atmosphere, in which such materials have both a macro and a micro effect, only the latter becomes important in water. In other words, added chemicals can cause water to be unacceptable for human use (a micro effect). However, there is no example where the addition of chemicals to water has a global impact (macro effect). Consequently, attention will be focused on the micro effects of water contamination.

Water is the most abundant substance on the earth's surface. The oceans cover approximately 71% of the planet, an observation that makes earth unique in the solar system. Of all the planets, this is the only one that is blue when observed from space. Besides the water in the oceans, the liquid is also found in glaciers, ice caps, lakes, streams, soils, and underground reservoirs.

Homes, industry, agriculture, and recreation all use water. The ubiquitous nature of water is the main reason that it becomes contaminated. Water pollution occurs when some substances (chemicals) or conditions (such as heat) so degrade the water that government standards for that use are violated. While some uses may be prohibited, the water may be suitable for other uses. For example, water that is too polluted to drink may be satisfactory for industrial use. On the other hand, water that is too polluted for swimming may also be unsuitable for fishing. The following chapter will discuss this valuable resource.

PROPERTIES OF WATER

Most chemical substances can be classified into groups with similar properties. Water, however, is in a class by itself. There is nothing similar. To understand the reason for this statement, the molecular structure as shown in

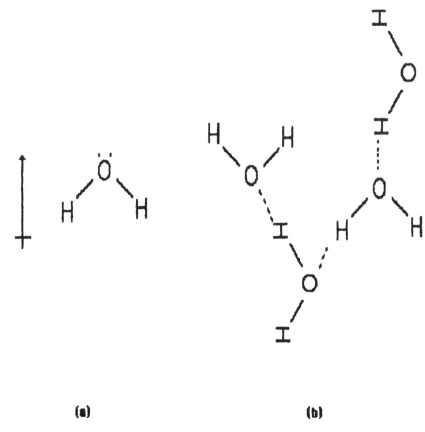

(a)   **(b)**

Figure 1.   **(a) Structural formula for water. (b) Illustrating the bonding
that can occur between water molecules. The arrow on the
left indicates the polarity with the arrow pointing in the negative direction.**

Figure 1 must be examined. The molecule has a bent shape with an angle of
105° between the bonds (Figure 1a). Oxygen is electronegative so that the
negative charges are concentrated on the oxygen atom leaving the hydrogen
side with a deficiency of electrons or slightly positive. The consequence of
this charge separation is that water molecules are bonded together in a network
as shown in Figure 1b. The network is called hydrogen bonding. The shape
and the bonding lead to some important characteristics. As a result, water
more than any other substance has a strong influence on the environment.
Three of these influences are discussed below.

1.   Table 1 shows materials similar to water that have much lower
     melting and boiling points. The reason for this striking difference

Table 1.   Melting and Boiling Points for Several Materials

Compound	Boiling point (°C)	Melting point (°C)
Water (H_2O)	100	0
Methanol (CH_3OH)	65	−93
Methane (CH_4)	−164	−183
Ammonia (NH_3)	−33	−78

is the strong hydrogen bonding in water that keeps ice solid up to 0°C. Ammonia and methane, on the other hand, do not possess such bonding. Consequently, very little energy is required to break the solid into liquid. Even the liquid vaporizes at a very low temperature. Methanol is intermediate in the strength of its hydrogen bonding system. This is illustrated in the temperature where the solid changes to a liquid (−93°C) and the liquid is converted to a gas (65°C). Without hydrogen bonding, water would exist as a vapor at room temperature, and life as we know it would be impossible.

2. When ice does melt, the results are equally remarkable. Melting is a process in which the regular crystal structure is broken and the liberated molecules can move around freely. For most substances, the breakup causes an expansion, resulting in a liquid less dense than the solid. The consequence is that most liquids upon freezing become more dense and sink to the bottom. Water is just the opposite. When the crystal structure of ice is broken, the hydrogen bonding causes the molecules in the liquid to pack closer together. This makes the liquid denser than the solid so that ice floats. If this were not the case, a body of water would freeze from the bottom up and aquatic life would not exist in the part of the hemisphere where the temperature falls to 0°C.

3. The third characteristic is closely associated with the above. To break the hydrogen bonds in water, large amounts of energy are required. The transformations are illustrated in Equation 1. Thus, 334.7 J/g are required to convert ice to water at 0°C and 2259 J/g to convert liquid water to steam at the boiling point. Compare the boiling points of water with other liquids in Table 1. In those materials that lack hydrogen bonding, the energy required to change a liquid to a gas is much less.

$$H_2O \text{ (solid)} \quad \rightarrow \quad H_2O \text{ (liquid) 334.7 J/g} \qquad (1)$$
$$H_2O \text{ (liquid)} \quad \rightarrow \quad H_2O \text{ (gas) 2259 J/g}$$

Table 2. Specific Heats for Various Materials

Material	Specific heat (J/g/°C)
Ethanol	10.29
Water	4.184
Carbon tetrachloride	3.59
Methanol	2.63
Benzene	1.88
Iron	0.46
Zinc	0.38

This type of reaction in which the system (ice) loses heat to the surroundings is called endothermic. The reverse process, i.e., freezing of water, liberates 334.2 J/g and is termed exothermic.

In comparing the amount of heat required to make these transformations from one material to another, the term heat capacity is used. This is the quantity of heat necessary to raise the temperature of the sample 1°C. A related term is the specific heat or the amount of heat required to raise the temperature of 1 g by 1°C. Table 2 shows that, gram for gram, water has the capacity for absorbing large amounts of heat, a material with a high heat capacity. By contrast, very little heat is required to raise the temperature of iron or zinc. This characteristic gives to water bodies such as lakes an ability to moderate the temperature of the surrounding land. By storing heat in the summer they make the land area cooler. Similarly, by slowly liberating the heat in the fall before freezing, they cause the nearby land temperature to rise.

GLOBAL WATER BALANCE

Just as there was a natural cycle for carbon (Chapter 5), so there is a hydrological cycle. The amount of water on the earth is a constant in continuous flux among the three states: ice, liquid, and vapor. Figure 2 shows a simplified cycle where the units are in exagrams (Exa = 10^{18}; thus, Eg is 10^{18} grams or 1000 km³) (Strahler and Strahler, 1973).

The total evaporation (455 + 62) 517 Eg/year represents the total amount of water that must be returned to the liquid or solid state in a year. Notice that more water is received by the land, 108 Eg/year, than is returned by evaporation. The difference (46 Eg) referred to as runoff, flows over and under the soil and eventually reaches the ocean.

Great inequalities exist in global amounts of water stored in these compartments. Table 3 gives the distribution of the total water (1.35×10^{18} m³ or

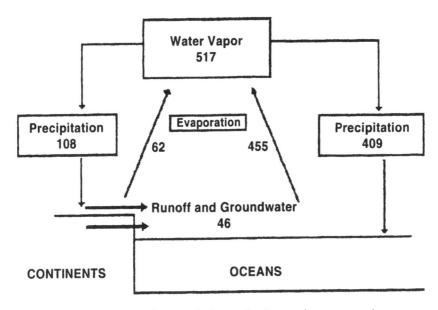

Figure 2.   Global water balance (units are in exagrams).

1.35×10^{21} L) (Neely, 1980). How long does it take for water in a given part of the earth to renew itself? The average time that a water molecule spends in any one location is called the residence time. The atmosphere is the compartment within which water is in a rapid flux (0.027 years). This should be compared to oceans and ice caps where the flux is measured in thousands of years.

The 3000-year residence time for water in the ocean is an average. More realistically, the ocean can be divided into two layers, the surface and a deep layer (below 100 m). The residence time for water in the surface layer is short (100 to 150 years), while the average time spent in the deep ocean ranges

Table 3.   Volume of Water and Average Residence Time for Each Compartment

Compartment	Volume $(1 \times 10^{15}$ L)	% of Total	Average residence time (years)
Ocean	1.3×10^6	97	3000
Ice caps and glaciers	28,900	2	8000
Groundwater	9,100	0.6	500
Lakes and rivers	230	0.016	7[a]
Atmosphere	13		0.027

[a]Rivers have a much shorter residence time (0.03 years).

(100 to 150 years), while the average time spent in the deep ocean ranges from 30,000 to 40,000 years. Similarly, the residence time for groundwater will vary by orders of magnitude, depending on the depth or inaccessibility of the reservoir. Pollution of these types of remote waters is not easily reversed. Consequently, there is great concern over the contamination of the water compartments with long residence times, such as aquifers and the deep parts of the oceans.

A brief description of the major compartments will be given. While the ice caps and glaciers are large storage reservoirs for fresh water, they are essentially isolated from sources of pollution and therefore unimportant from an environmental risk point of view; they will not be discussed further.

WATER COMPARTMENTS

The ability to identify compartments and assess their size is important from a risk assessment perspective. Combining the volumes with a value of how much chemical was added allows for decisions to be made on concentrations. Matching the concentrations with toxicity is the first step in evaluating the potential hazard. The investigation of how chemicals move through the many possible environmental compartments is referred to as a mass balance study. This topic will be addressed in the next chapter.

The Oceans

Table 4 lists the major definable areas of the oceans. Similar to the ITCZ zone (Chapter 5), ocean currents have created two major compartments north

Table 4.   Areas of the Major Ocean Bodies

Name	Area (millions of km^2)	Average depth (m)
Pacific	165	4267
Atlantic	81.3	3926
Indian	73	3962
Arctic	14.2	1280
Mediterranean	2.8	1371
Bering Sea	2.3	
Caribbean Sea	1.9	2560
Gulf of Mexico	1.8	1432
Hudson Bay	1.2	1432
Total	343.50	
Area north of Equator	154	
Area south of Equator	209	

and south of the equator. As indicated earlier, a surface volume in the ocean has been identified in which most of the pollutants are found. This volume is insulated from the deep portion. Consequently, the latter does not come into play as far as most contamination problems are concerned. Accordingly, attention will be focused on the top layer. To perform a mass balance study some knowledge of the depth of this layer is needed. Investigation of this topic has only begun. By measuring CCL_3F, a chlorofluorocarbon (see the case study on ozone in Chapter 9) in the North Atlantic (Lovelock, 1971), the depth of the layer was estimated as 130 m. In similar studies using carbon tetrachloride (Lovelock, 1971) a depth (referred to as the mixing depth) of 50 m was estimated. A more thorough investigation (Hammond, 1977) using tritium analysis showed a variable mixing depth with 100 m being a reasonable average. Using 100 m, a mass balance study was applied to the input data for chlorofluorocarbon (Neely, 1978). The estimated concentrations in the ocean were in reasonable agreement with the values as measured by monitoring (Lovelock, 1971).

Lakes

On a volume basis, the water in the rivers and lakes represents a very small percentage of the total. However, there is no question that this particular water is very important as far as our lifestyle is concerned. The reasons are many, and only two will be given. First, and most obvious, fresh water is necessary to sustain life. Second, the major rivers that crisscross the country represent an important artery for the economical movement of goods from production to consumers. Rivers and lakes become a local problem as contrasted with oceans. Table 5 summarizes the key data for the Great Lakes that represent 20% of the world's supply of fresh liquid water and Lake Baikal (in Russia) which depicts another 20%. Because of their low flow rate and large volume, the flushing time (residence time) for lakes is 1 to 100 years compared to

Table 5. Physical Parameters Relating to the Great Lakes

Lake	Area (km^2)	Volume (km^3)	Residence time (years)
Superior	8.17×10^4	1.22×10^4	106
Michigan	5.77×10^4	4.9×10^3	66
Huron	5.93×10^4	3.53×10^3	21
Erie	2.55×10^4	4.80×10^2	2.3
Ontario	1.96×10^4	1.60×10^3	6.9
Baikal[a]	5.08×10^4	2.49×10^4	

[a]Greatest depth is reported to be 1.6 km.

weeks for rivers. The long residence time means that the water in some lakes remains essentially constant for at least a year. In the mid-latitudes this creates a situation in which the temperature of the water varies with the depth. During the warm summer months the upper layer (epilimnion, Greek for "overlake") is heated and floats over the cold dense layer of the bottom or hypolimnion (Greek for "underlake"). The layers are separated by a thin region called the thermocline. The location of the thermocline will vary with atmospheric temperature. For example, as fall approaches the epilimnion cools and the thermocline moves to the surface. At the time of uniform temperature the lake is said to have "turned over" and in the process becomes completely mixed. This movement exposes the bottom layer to the surface, enabling the oxygen in the water to be replenished. For persistent chemicals whose degradation and dissipation reactions are very low, the mixing in lakes can be considered uniform from top to bottom, once turnover has occurred. However, for easily degraded chemicals, some knowledge is needed concerning the location of the thermocline. This information can be used to estimate the volume of water in which the contaminant is dissolved, and ultimately the concentration of the contaminant.

Another natural characteristic of lakes is eutrophication (Greek for "well nourished"). This is a process of enrichment of the water with plant nutrients. A lake with a low supply of nutrients is said to be oligotrophic (Greek for "few sources of nutrients"). A eutrophic lake has a high algae population and a low dissolved oxygen content. The fish species tend to be carp and bullhead. Eutrophication is a natural process eventually leading to a situation in which the lake becomes completely filled (swamp) and then turns into a peat bog. The natural process takes thousands of years. A widespread pollution problem is caused by civilizations speeding up the eutrophication process by discharging an excess of nutrients into the lakes. This is called cultural eutrophication and can produce a eutrophic water body in decades as opposed to centuries. A classic example of cultural eutrophication is the eutrophication of Lake Erie by the discharging of phosphates, fertilizers, and sewage from the watershed serving this lake. Through regulation the process has been slowed and Lake Erie is showing signs of recovering from the heavy pollution load that occurred in earlier years (1960 to 1970).

Rivers

Estimating concentrations of contaminants in rivers is easy for a steady-state situation. In such a situation the only knowledge required is the volumetric flow rate of the river and the rate of discharge of the effluent into the stream. For example, assume a manufacturing plant is discharging 0.2 m³/second of water (7 ft³/second) into a river flowing at 30 m³/second (1060 ft³/second).

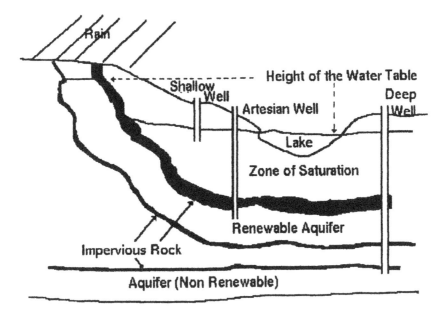

Figure 3. **Schematic illustrating the categories of groundwater and the three types of wells for obtaining water: shallow, artesian, and deep.**

The dilution factor will be

$$0.2/(30 + 0.2) = 0.007$$

By knowing the concentration of the chemical in the discharge stream an estimate of the concentration in the river may be obtained by multiplying the discharge concentration by 0.007.

Groundwater

There are two main categories of groundwater, renewable and nonrenewable. These are illustrated in Figure 3. Groundwater is renewable when the aquifer is open to the surface and can be replaced by rainfall. Three types of wells for obtaining the groundwater are shown in the diagram, two from the renewable zone and one from the nonrenewable zone. The first is the shallow well that penetrates the zone of saturation and requires a pump to bring the water to the surface. The second type is the artesian well. In this case the height of the water table is above the outlet, creating a pressure head which forces the water to the top creating a constant flowing well. The third type is the deep well that penetrates into the nonrenewable zone. Nonrenewable aquifers are not open to the surface and were created thousands of years ago during a period of heavy and frequent rainstorms. Once the water has been removed, it is gone. Similar to coal and oil, these water deposits have been called "fossil" water. Removal of deep groundwater is analogous to mining. The Ogallala aquifer under the central states of Nebraska, Colorado, Kansas, Oklahoma, and Texas is one of the world's largest reservoirs of fresh groundwater, and is an example of fossil water. The discovery of this deep aquifer changed the nature of the farming in this area. Before the discovery they practiced "dryland" farming. However, irrigation allowed for greater intensity of agriculture and increased the productivity and income of the farmers in the region. The one alarming feature is that, once the water is gone, farming must revert to the former condition of dryland farming. The challenge will be to see if these agricultural communities, living on a limited supply of water, can make the transition back to dryland farming or find another supply of water. At the present rate of use the Ogallala aquifer, which contains as much water as Lake Michigan, will be depleted within 40 years (Tyler, 1985). This is an example of leaving a major problem for the next generation to solve. Hopefully, society will be alerted to the challenge before the problem creates economic and environmental chaos.

COMPOSITION

Pure water, by definition, consists entirely of H_2O molecules. This would be rare, expensive, and not very tasty. The presence of salts such as carbonates give water the taste that has become acceptable. Water pollution occurs when the concentrations of added materials exceed certain defined standards as set by the states and the EPA. One useful classification of such constituents found in water is based on size.

1. *Suspended particles*—When the diameter is greater then 1 μm, the particles settle out under gravity. An example of water carrying a high level of suspended particles would be rain water running off a dirt road.
2. *Colloidal particles*—These particles are so small that the settling rate is insignificant. Water containing colloidal material will appear cloudy when viewed at right angles.
3. *Dissolved*—Finally there are substances that form true solutions and cannot be seen.

Natural water contains all three of the above, and the taste ranges from potable to poisonous and fresh to salty. The dissolved materials are the most difficult to control. Typically they are tasteless, colorless, and odorless, and some may be toxic.

The most important constituent found in water is oxygen. While the amount is small compared to air (8.4 mg/L in water: 270 mg/L in air), it is very critical for the support of aquatic life. There are two processes for generating oxygen in water. The first is by aeration, either mechanically with air pumps or naturally through movement such as that occurring in rapids and wave action. The second is by photosynthetic aquatic plants that produce oxygen in the presence of sunlight. Both processes are important for enriching water with oxygen.

In clear oligotrophic lakes the amount of biological activity is small; therefore, the depletion of oxygen is slow. The natural replacement from air is sufficient to maintain a proper level. Lake Superior is an example of such a system. As lakes or streams gain in nutrients, i.e., become more productive, the biological activity increases and the demand for oxygen increases. When the rate of depletion is greater than the rate of replacement, the body of water becomes anaerobic (lack of oxygen). This produces a condition where foul-smelling odors become very prevalent. A term used to describe the amount of oxygen consumed is the biochemical oxygen demand or BOD. This is also called the index of pollution. A high BOD means that the water requires the addition of oxygen or the water will become anaerobic. In most river systems, the flowing water usually causes a quick replacement of oxygen. Figure 4 illustrates the events that can occur as a discharge containing a high level of BOD is added to a river.

As waste is added, the degradation of the material consumes the dissolved oxygen and the concentration in the stream falls. If enough material has been added to cause the oxygen sag to last for several hours, then a fish kill will be observed. Usually oxygen is replaced quickly so that the river recovers with no lasting effect. Trouble results when discharges are too close together and a discharge is added before the recovery of the stream from a previous insult. In this situation the river is acting as a waste disposal system and is unsuitable for other uses. Government regulations are gradually forcing communities to operate more responsibly. This is another example of "mutual coercion" as described in Chapter 1 (Hardin, 1968). The suggestion has been made that, if each community were to withdraw drinking water downstream of their discharge pipe rather than upstream, water quality would improve dramatically!

The major water pollutants fall into one or more of the following categories.

1. Oxygen-demanding material—This category includes domestic sewage and industrial waste. If the water systems are overloaded with such materials, the resulting explosion of microbial growth uses the oxygen to the point where most fish would not survive. In such a case, the body of water becomes anaerobic.
2. Disease-causing agents such as bacteria and viruses.
3. Inorganic chemicals.

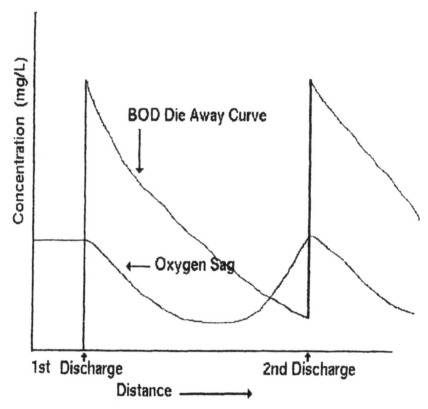

Figure 4.   **Schematic illustrating the events that occur in a stream as a high load of organic material is added.**

4. Synthetic organics such as plastics, detergents, pesticides, etc.
5. Sediments from soil erosion.
6. Radioactive materials.
7. Heat from industrial and electric power plant cooling water.

CONCLUSION

Potable water is a priceless commodity and one that needs to be available to everyone. How this resource is managed will become a key issue in the years ahead. For example, a recent newspaper article[1] discussed the skyrocketing water bills in cities such as Los Angeles and Boston. The huge increases are a result of the costs involved in complying with the clean water laws. In

---

[1] *The Wall Street Journal*, Jan. 15, 1992.

Boston, the problem is becoming most severe as the cost of cleaning up Boston Harbor has been phased into the water bills. By the year 2000 the average annual water bill in Boston is expected to rise from the present $500 to about $1600. People are beginning to ask questions about the value of environmental projects related to water, especially those groups of people who can't afford to pay their water bills.

These and other issues will form the basis for serious debate now and in the future on the subject of costs vs. benefits.[2] The discussions will revolve around questions such as "How clean is clean?" and ". . . are the benefits of the increased cleanliness in terms of improved health worth the costs?" This chapter has established a beginning base to formulate questions to help find the answers.

---

[2] In any discussions of such a topic it must be accepted that every person should have affordable access to clean water and clean air.

7   EXPOSURE ASSESSMENT

INTRODUCTION

Chapters 3 and 4 discussed some of the toxicological effects of chemicals, while Chapters 5 and 6 described the properties of the two main compartments through which exposure occurs. The next task is to make an assessment of the risk that the biological systems in these compartments face by being exposed to the contaminant. There are two methods for performing such analysis, both of which involve the matching of concentrations with the toxicological response.

The first, called monitoring, deals with analyzing samples taken from different environmental compartments for the suspected chemical. As the method implies, it can only be used if the material has been released into the environment either intentionally (pesticides) or accidentally (industrial chemicals) in sufficient amounts to be detected analytically.

The second method involves the use of mathematical models. Here the objective is to simulate an environment that reasonably reflects the planned use of the chemical. By estimating the concentrations with these models, conclusions can be made regarding the hazard that the ecosystem might experience. The next sections describe these methods.

MONITORING

If the chemical has been produced for many years, there is a distinct possibility that the material may have entered the environment and probably can be detected analytically. By initiating a monitoring program, data can be gathered for establishing the levels of the chemical in the biosphere. These levels are then matched with toxicological data and decisions made regarding the potential impact. It is entirely possible that the detected levels may not be great enough to produce a biological response. However, before concluding there is no risk, the production history of the chemical must be examined.

Table 1.   Domestic Uses of the Polychlorinated Biphenyls (PCBs)

Systems	Uses
Electrical (closed)	Transformers, capacitors, insulating cooling applications
Nominally closed[a]	Hydraulic fluids, heat transfer, fluids, lubricants
Open-ended uses[a]	Plasticizer, surface coatings, adhesives, pesticide extenders, carbonless copy paper, dyes

[a]After the voluntary curtailment in 1971 these uses dropped to zero.

For example, if the demand for the material is growing, then in a few years the additional burden might create a hazard, and efforts should be made to reduce the input. Two examples will illustrate this aspect of exposure assessment in greater detail.

Polychlorinated Biphenyls (PCBs)

Beginning in the 1930s a fluid formed by chlorinating biphenyl was used for a variety of purposes ranging from closed-end systems such as electrical capacitors to open-ended systems such as insecticide extenders (Table 1).[1] The fluids were found to be very useful because of their lack of reactivity, low degree of flammability, and the ability to act as solvents. The demand caused an exponential growth[2] in their sales and distribution. By 1975 over 20×10^8 kg of PCBs had been produced worldwide. In the U.S. the total was estimated at 6.1×10^8 kg (Beeton, 1979). If the initial production in 1930 was about 1,000,000 kg, then an average growth rate of about 9.5% would be required to yield the cumulative total. By any measure this was a very rapid growth, one that was doubling every 7 years. As of 1975, 3×10^8 kg or 60% was thought to be still in service. The remainder was distributed as follows (Beeton, 1979):

Mobile PCBs circulating in the environment	68.2×10^6 kg
Degraded in incinerators	25×10^6 kg
Landfills, dump sites, etc.	130×10^6 kg

[1] Open ended refers to uses where the product is free to move directly into the environment. Closed ended are uses where the product is contained, such as in a transformer, and is not free to move into the environment.

[2] Appendix II describes the characteristics of the exponential curve.

Table 2.   The Geographical Distribution of PCBs in Sediments from January 1971 to June 1972 in the U.S.

State	Median concentration (μg/kg)
Arkansas (23)	60
California (13)	85
Georgia (12)	300
Maryland (11)	30
New Jersey (12)	20
Texas (293)	80

Note: The numbers in parentheses indicate the number of samples.

Extensive monitoring showed that the mobile PCB had become widely distributed. Table 2 is just one of many examples showing the geographical distribution of PCBs in the bottom sediments of the waterways throughout the U.S.

The finding of PCBs in the Great Lakes is another illustration of the contamination of the nation's waters. PCBs in the surface waters of the lakes were found to range from a high of 31 ng/L in Lake Michigan to 1 ng/L in Lake Ontario and Lake Superior.[3]

With this widespread distribution of PCB, investigators were beginning to discover that the presence of these chlorinated hydrocarbons was having an impact on biological systems, for example:

> Michigan ranchers in the early 1960s began noticing reproductive complications and excessive birth mortality in their mink herds (Beeton, 1979). By 1967 an acute problem was in evidence resulting in an unprecedented 80% newborn mortality. Subsequent investigations suggested a relationship between the birth defects and the percentage of coho salmon in the mother's diet. Further experiments showed that diets of 30% salmon produced the problem. Other species of Great Lakes fish caused similar problems. Shortly after that the presence of high concentrations of PCBs in the fish was implicated as the main cause.

During the early part of 1970, many fish taken from Lake Michigan were found to have PCB levels in the edible portion of the fish in concentrations ranging from 10 to 20 μg/g (ppm). Because of these levels the Food and Drug Administration (FDA, 1979) set tolerance levels of 2 μg/g for fish sold commercially.[4] With all of this activity the major producer, Monsanto Co., decided to voluntarily curtail the sale of PCBs for all open-ended and nominally open-ended uses (see Table 1).

---

[3] A low value of 0.8 ng/L in Lake Superior is used as a conservative estimate for freshwater background levels of PCB (Beeton, 1979). This is based on the assumption that this lake is largely pristine water that has been insulated from industrial pollution.

[4] This number is driven by economics, as no official risk assessment based on toxicological data has been performed.

Table 3.   Monitoring Results for CCl_3F in the Troposphere

| Year | Concentration | |
	pmol/mol	μg/m³
1971	50	0.285
1972	60	0.342
1974	70	0.399

As more PCBs were found, pressure began to mount to purge the ecosystem of the chemical. Eventually, the EPA placed a ban on the production and importation of all chlorinated biphenyls. Recent monitoring shows that the ban is meeting its objective as the levels of PCB contamination are decreasing.[5]

Chlorinated Solvents

During the 1950s and 1960s there were several volatile chlorinated hydrocarbons manufactured and sold for different applications such as cleaning solvents, aerosol propellants, refrigerants, etc. The majority of these uses were open-ended and, similar to the PCBs, the chemicals ended up in the environment. Because of their volatility the main compartment where they were found was the atmosphere. The two classes that will be described are the Freons[6] and methylchloroform.

Freons

The Freons are materials that were used as refrigerants and aerosol propellants (see the case study on ozone in Chapter 9). From 1957 to 1973, use of CCl_3F (Freon 11) grew at a rate of 18% a year, reaching a production figure of 3.67×10^8 kg/year in 1973 (Neely, 1980). The companion material CCl_2F_2 (Freon 12) had a similar growth pattern reaching a production figure of 4.4 $\times 10^8$ kg/year in 1973.

At the time there was no environmental concern since the materials were unreactive and had no significant toxicity. The most important contribution they made to environmental science was as markers for measuring and discovering how chemicals were distributed in the troposphere. Lovelock was one of the key investigators in this area and some of his monitoring results are summarized in Table 3 (Neely, 1977). During the same interval, scientists from the University of California (Molina and Rowland, 1974) raised questions

[5] The *Midland Daily News* (Midland, MI) for Tuesday, April 7, 1992, reported on a conference in Lansing, MI which indicated that the levels of PCBs had dropped 90% in some of the state's lakes.
[6] Freons are a registered trademark of the DuPont Co.

Table 4.   Monitoring Data for Methylchloroform in the Troposphere

| Year | Concentration | |
	pmol/mol	µg/m³
1972	25–50	0.14–0.28
1974	40–60	0.23–0.34
1975	60–75	0.34–0.43
1976	70–90	0.40–0.51
1977	85–95	0.48–0.54

about the ultimate sink for these volatile agents. After much study the conclusion was reached that the chlorofluorocarbons would eventually reach the stratosphere.[7] Here they would undergo decomposition with the resultant destruction of the ozone layer. The combination of an exponential growth and the destruction of this important layer caused a sharp curtailment in the use of these chemicals (see Chapter 9).

Methylchloroform

The Freons in the atmosphere were measured with an instrument known as a gas chromatograph. This instrument separated the mixture into individual peaks based on their volatility. In examining the air samples, the investigators found many peaks other than Freons. A challenging task was to identify these unknown entities. One peak in the chromatogram was eventually identified as methylchloroform (CH_3CCl_3). This is a solvent that had found wide use as an industrial degreasing agent, an open-ended use that allowed escape to the atmosphere. Further study revealed that this material had also been enjoying an exponential growth for several years. From 1960 to 1975 the annual growth rate had been about 16 to 17% reaching an annual production figure worldwide of 3.7×10^8 kg (825 million pounds) in 1975 (Neely and Plonka, 1978). Monitoring data performed by Lovelock and summarized in Table 4 (Neely and Plonka, 1978) displayed the general increase in the atmospheric concentration. With these observations speculation was raised as to the environmental significance of this chlorinated solvent. Between 1975 and 1986, this issue became a subject for intense debate between the producers of methylchloroform and the environmentalists. The latter group argued that methylchloroform was similar to the Freons and would have the same impact on the ozone layer. The producers, on the other hand, cited evidence that the halogenated hydrocarbons would photodegrade in the troposphere. If methylchloroform

---

[7] Appendix II describes a simple mass-balance study which was the basis for concluding that the Freons would eventually end up in the stratosphere.

turned out to have a stratospheric impact, the lower atmosphere could be purged very quickly, and so alleviate the problem before it became acute. These arguments became quite mute in the later part of the 1980s with the identification of the "greenhouse effect" (Chapter 5). As will be recalled, this is an effect associated with possible global warming. Methylchloroform was identified along with several other gases such as CO_2 and the Freons as one of the culprits and pressure began to mount to end production. On April 10, 1992, the Dow Chemical Company, a major manufacturer, announced that it would phase out production by December 31, 1995.[8]

Conclusion

Monitoring is a useful technique for following the environmental distribution of a chemical. By matching these concentrations with toxicological or other data, decisions can be made as to the potential problems. Unfortunately, by the time the material can be detected, the production must have had a sizeable growth. In such a case, if there is a potential hazard several years might be required to lower the concentration to a safe level. One general conclusion can be made from these and other examples. If a product has reached a constant rate of growth and residues can be detected in the environment, there is a high probability that a problem of some kind will emerge.[9] Nature does not like exponential growth and some event will surely occur to cause a flattening into what mathematicians call a logistic pattern.

Such a curve is shown in Figure 1. This is the form taken by normal biological growth as shown by humans, trees, and other biological organisms. In all cases there is a fast initial phase followed by plateau or leveling off. The differential equation for such growth is given in Equation 1.

$$\frac{dA}{dt} = (a - bA_0) A_0 \tag{1}$$

where

a = growth constant
b = damping factor, a function of mass and time
t = time
A = amount at time t
A_0 = amount at time = 0 (initial production)

---

[8] *Chemical & Engineering News*, p. 7, April 20, 1992.
[9] Appendix II describes the characteristics of the exponential curve.

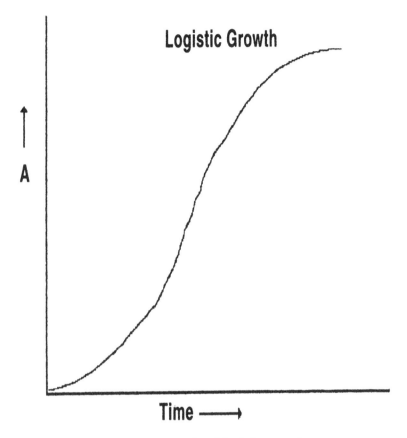

Figure 1.   An example of logistic growth.

If b is small such that a > bA, Equation 1 reduces to the simple exponential curve (Equation 2). Integration of Equation 1 yields Equation 3.

$$\frac{dA}{dt} = aA_0 \tag{2}$$

$$A = \frac{aA_0}{bA_0 + (a + bA_0) \exp{(-at)}} \tag{3}$$

At steady state, A approaches a/b. When production of a chemical reaches exponential growth, watch for the appearance of a damping factor (b in Equation 1) that will cause the growth to become logistic. Without such a factor, the continued production might cause an environmental problem.

As can be seen monitoring is useful only after the fact. What is needed is a method to predict the consequences of a course of action before it happens. The next section describes such an approach.

MODELING

Environmental modeling is an attempt to describe mathematically how chemicals move and distribute themselves in the biosphere. This can be accomplished since chemicals are under the influence of natural laws as opposed to random processes. More specifically, chemicals are subject to the laws of conservation of mass and behave in a predictable manner. An excellent book on the subject is *Multimedia Environmental Models* (Mackay, 1991). There are two types of models that will be presented. The first is concerned with site specific situations while the second is more general in nature. Illustrations will be presented, where the emphasis will be on the results with no attempt at explaining the mathematics. For more detailed information, the reader is referred to the original literature.

Site-Specific Modeling

There are times when answers are needed regarding the toxicological significance of concentrations arising from the discharges of chemicals into the environment. These incidents can take the form of

1. Spills into rivers, lakes, or ponds
2. Spills on the ground
3. Discharge from stack gases at a manufacturing site

Often answers are needed before a monitoring program can be established, since the crisis will be over before the first sample can be taken. In other cases the concentrations may be so low as to be undetected. Readily available computer programs (Neely, 1992) can analyze the situation to allow action to be taken to minimize the hazard.

One example will illustrate the technique as it applies to estimating concentrations in stack gases. The hypothetical problem to be discussed is concerned with the exposure to individuals near a chemical plant. A chemical with a permissible exposure level (PEL) of 0.5 nmol/mol (ppb) is suspected of being in the plume emitted from the stack. For this example, the source strength of the chemical in the stack gas was 0.14 g/second. Equation 4 can then be used to estimate the concentration downwind from the stack (Turner, 1970).

$$C = \frac{Q}{\Pi \, \sigma_y \sigma_z U} \exp\left[-\frac{1}{2}\left(\frac{H}{\sigma_z}\right)^2 \right] \tag{4}$$

**Table 5.   Computer Calculations of Chemical Concentrations
Downwind from Stacks**

| Distance (km) | Concentration (nmol/mol) from a stack height | |
	100 m	20 m
0.2	3.5×10^{-5}	1.3
0.4	0.0025	0.03
0.6	0.0069	0.02
0.8	0.0083	
1.0	0.0081	0.012
1.2	0.0074	
1.4	0.0070	0.007
1.6	0.0067	
1.8	0.006	
2.0	0.005	0.003

Note: Concentrations are shown for the chemical downwind from two stacks with effective heights of 100 and 20 m. The source strength was 0.14 g/second with a wind speed of 4 m/second during daylight hours with a temperature of 25°C.

where

C = concentration at ground level and in the center line of the plume

σ = the standard deviation of the plume concentration distributed in the horizontal (y) and the vertical (z) direction

H = the effective stack height

U = wind velocity (m/s)

Q = source strength of chemical in g/second

The standard deviations are a function of the climatic conditions, and graphs and equations have been developed to take such items into account (Turner, 1970). The equations were solved using the computer package (Neely, 1992).

Table 5 shows the results from two computer runs. As the stack height is increased to 100 m, the concentration near the source is negligible (Figure 2). At 800 m downwind of the stacks, the concentration reaches a maximum and then decreases. Only with a short stack height (20 m) and within 200 m of the source does the concentration exceed the PEL of 0.5 nmol/mol. By using this simple approach, a decision can be made as to the height of stack that is necessary to minimize the exposure to the chemical. Alternatively, the decision could be made to reduce the source strength of the material to a level where the PEL will not be exceeded from the 20-m stack.

Global Distribution

The second type of modeling is concerned with a gradually increasing amount of a chemical released into the world over a period of many years.

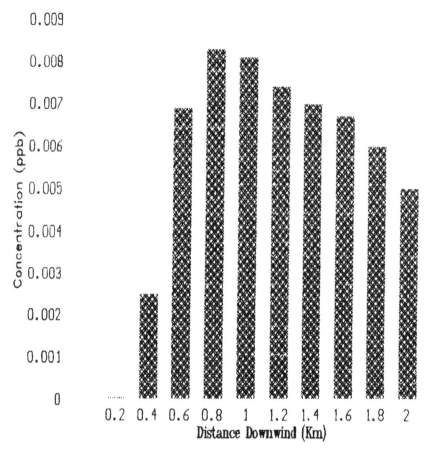

Figure 2.   Downwind concentrations from a 100-m stack.

What concentrations can be expected from the anticipated production or load and where will the maximum concentrations be located? Answers to these questions need to be found before production reaches the exponential growth phase. There have been many models designed to address such situations (Mackay, 1979, 1991; Mackay et al., 1985; Neely et al., 1982). Figure 3 shows the basic model, called the Unit World, illustrating the major compartments.[10] The scaling is such that 510 million units are equivalent to the world (Neely, 1985). Table 6 shows the volumes and physical properties. The model depicts a world where the annual rainfall is 0.7 m (30 in.) with an ambient temperature of 25°C. The slope of the ground has been set at 20% to exaggerate the amount of chemical transported to the water compartment by runoff. The ground itself represents a loam-type soil with an organic content of 2% and a water content

---

[10] This is a type of mass balance study described in Appendix II.

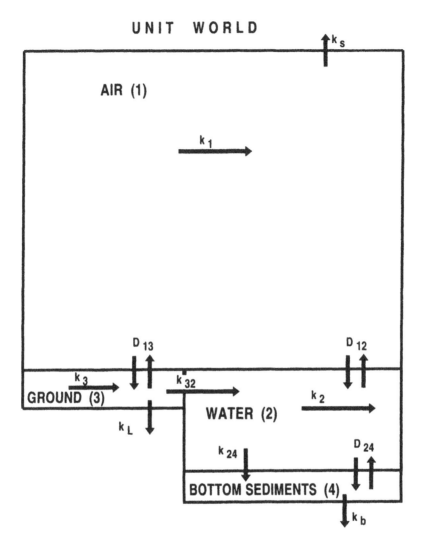

Figure 3.   Schematic representation of the Unit World. The double-headed arrows represent diffusion processes. The subscripts refer to the compartment numbers.

of 20%. A maximum water content of 100% and a porosity of 88% characterizes the bottom sediments. Once a chemical has been added to this model it is subjected to the dispersion forces indicated by the double-headed arrows in Figure 3. In addition, there are several processes that act in only one direction.

1.   Runoff (k_{32}) moves chemical absorbed to soil particles from the ground to the water. The amount of chemical transferred in this

Table 6.   Environmental Parameters for the Unit World

Compartment	Volume (m³)	Area (m²)	Depth (m)
Air	6×10^9		
Water[a]	7×10^6	7×10^5	10
Ground	4.5×10^4	3×10^5	0.15
Bottom sediments	2.1×10^4	7×10^5	0.03
Ground			
Bulk density	1.5×10^6 g/m³		
Water content	0.2		
Porosity	0.4		
Fraction organic	0.02		
Air content	0.2		
Slope	20%		
Bottom sediments			
Porosity	0.88		
Water content	1.0		
Exit to stratosphere	4.1×10^{-5}/day		
Burial constant	2×10^{-4}/day		
Suspended sediments	5 g/m³		
Average rainfall	0.7 m/year		
Average temperature	25°C		

[a]Water is 70% of the surface area of the world.

    manner is controlled by the slope, the extent of rainfall, and the organic content of the soil.

2.   Degradation processes (k_3, k_2, k_1).

Finally, to achieve a steady-state situation for persistent chemicals such as DDT[11] or the chlorofluorocarbons, there must be natural mechanisms for removing the material from the system. In the present model there are three such mechanisms:

3.   Exit to the stratosphere where the rate constant is based on the many mass balance studies that have been performed on the chloro-fluorocarbons (see, for example, Neely, 1977). In the Unit World persistent chemicals have a half-life in the troposphere of 46 years.

4.   Removal from the active sediment layer on the bottom of the water column to the deep inactive layer of sediment (Neely and Oliver, 1986). This process has a half-life of 9.5 years.

5.   Leaching through the ground beyond the 2-m depth (Neely and Oliver, 1986).

---

[11] DDT: Dichloro diphenyl trichloromethane, an insecticide used extensively in the 1950s and 1960s.

Table 7. Results of Adding DDT to the Unit World

Properties of DDT

Molecular weight 354

Vapor pressure 2.6 × 10^{-9} Pa

Water solubility 0.003 mg/L

Log Kow 6.00

Degradation constants
  were set equal to 0

Loading rate
  2.4 mmol/day
  75% to the ground
  25% to the water

Results

Residence time 134 years

% in the water 0.72

% in the air 0.76

% in the ground 93.55

% in the bottom sediments 4.98

Concentration in water .042 µg/kg

Concentration in air .065 µg/m^3

Concentration in ground 864 µg/kg

Concentration in the bottom sediments 98 µg/kg

All chemicals added to the model are subjected to these five processes. The final, and sometimes the most difficult, parameter to establish is a value for the loading rate in units of mass/time. For chemicals, such as pesticides, where the total production is designed to be added to the environment, the loading rate is related to the production. For other types of industrial chemicals, decisions need to be made regarding the fraction of the production that might enter the biosphere.

To partially validate the model as a technique for predicting environmental concentrations, the insecticide DDT will be examined (Table 7). Records show that worldwide production of the chemical reached a level of 1.6 × 10^8 kg (350 million pounds) in the early 1970s (Neely, 1985). Dividing the value by 510 million and converting mass to moles the equivalent loading rate for the model becomes 2.4 mmol/day. Since the chemical is a pesticide it will be further assumed that the entire production entered the environment—75% to the ground and 25% to the water. The outcome is shown in Table 7, where several things should be noted.

1. The results are steady-state values. In other words, at steady state, the amount leaving the model equals the amount entering. The length of time required to reach this point is a measure of the

persistence of the chemical. The long residence time is characteristic of a persistent material. For DDT the residence time is 134 years. Since DDT is persistent, 134 years becomes a benchmark to gauge persistence.

2. The numbers are not to be taken too literally; however, they do show a trend as to what might happen long-term with the continued input of DDT.

3. By comparing the estimated water concentration of 42 $\mu g/m^3$ with the water quality criteria number of 1 $\mu g/m^3$ (Costle, 1980) the exposure from DDT would become too great with continued long-term heavy input of the insecticide.

4. In the case of DDT the final regulatory action was a ban on the use that resulted in a lowering of the loading rate. This caused the exit rates to slowly begin the task of clearing the environmental compartments of the pesticide.

5. The results also suggest why there was so much effort applied to finding routes of degradation to supplement the three mechanisms of leaching, burial, and exit to the stratosphere for clearing the environment. If such a mechanism could have been found, then the continued use of DDT might have been allowed.

The evaluation of PCB using the data in Table 8 will be conducted in a manner similar to the above. This is accomplished by dividing the worldwide production figure for PCB (20×10^8 kg/year) by 510 million (Beeton, 1979). The equivalent production figure for the model becomes 3.68 nmol/day. Of this amount, 60% was in closed-end use and therefore insulated from the environment, leaving about 1.46 mmol/day as the loading rate for the model. Using the data supplied by the National Academy of Sciences (Beeton, 1979), 30% (0.44 mmol) is circulating and will be added to the water compartment and 60% (0.92 mmol) is located in landfills and will be added to the ground compartment. The remainder of the 1.46 mmol will be assumed lost by incineration.

By comparing the concentrations in water (42 ng/L) and bottom sediments (98 $\mu g/kg$) with the monitoring results cited earlier, the conclusion is reached that the model does describe what happened in the real world. This type of validation lends credibility to using this technique for assessment of environmental concentrations. For example, if such an analysis had been used in the 1950s there would have been recognition that continued use of the many open-ended applications of PCBs would create a hazard. This was the time that a decision to restrict the sales of PCBs to closed-end uses should have been made. Making the early decision would have prevented the environmental insult from PCB, and the use of the material as a dielectric and capacitor fluid might have been continued.

Table 8. **Results of Adding PCB to the Unit World**

Properties of PCB[a]

Molecular weight 292
Vapor pressure 0.0001 Pa
Water solubility 0.068 mg/L
Log Kow 6.44
Degradation constants
  set equal to zero
Loading rate 1.36 mmol /day
           66% to the water
           34% to the ground

Results

Residence time 159 years
% in the water 0.34
% in the air 0.05
% in the ground 93.05
% in the bottom sediments 6.56
Concentration in water 0.001 μg/kg
Concentration in air 1.78×10^{-6} nmol/mol
Concentration in ground 434 μg/kg
Concentration in the bottom sediments 65.6 μg/kg

[a]The properties of the 2,4,2′,4′-tetrachlorobiphenyl were used as a representative of the PCB isomer (Mackay et al., 1983).

Conclusion

There is no doubt that modeling is a very useful technique for assessing the potential environmental exposure of a new chemical. By estimating a value for the loading rate and choosing realistic numbers for the volumes of the compartments, reasonable predictions can be made for the expected concentrations. Such a procedure can be employed in the early planning stage of a new product and help prevent future environmental problems from occurring.

8   RISK BENEFIT ANALYSIS

INTRODUCTION

The previous chapters have discussed both hazard and exposure. Together these elements constitute risk. The other part of the equation is the benefit side. As mentioned earlier, the issues that are the focus of this book are situations where the risk and benefit are distributed unevenly among the members of society. In many ways the risk is much easier to deal with than the benefit. One problem is deciding who benefits. An advantage for one person may be quite different for another. Resolution of this difficult question is in the arena of public policy. Wessel, in his 1980 book *Science and the Conscience* refers to these issues as socioscientific. The term implies that the problems have a scientific element and a direct involvement with the public. A partial list of such issues (some are discussed in greater detail in the case studies in Chapter 9) includes the following:

1. *Supersonic transport and aerosol can.* Do these industries cause a destruction of the stratospheric ozone and what is the impact on society? Aerosol cans were introduced because of their ability to deliver a uniform coating on whatever object was being sprayed. At that time no one believed that the propellant might be harmful to the stratospheric ozone located 10 miles above the surface of the earth. The belief is that a similar destruction of the ozone layer is caused by supersonic air travel. The direct benefits derived by the user, i.e., ease of spraying paint or faster travel by the users of the supersonic plane, may cause a negative impact on society as a whole. Whose benefit comes first? How should these questions be resolved?

2. *Alar on apples.* When Alar is sprayed on apples the fruit stays on the trees longer and therefore ripens more fully. This allows increased production of a fruit that is more pleasing from an aesthetic point of view, i.e., "everyone loves a red apple." The

alleged risk according to the Natural Resources Defense Fund and aired by CBS's "60 Minutes" is that spraying with Alar produces an apple contaminated with a carcinogen. The question of risk vs. benefit can only be answered through an analysis of the data.

3. *Dioxin contamination*. When a manufacturing plant produces a variety of products and provides employment to thousands of people, the benefit of the company is obvious. However, when that same plant unwittingly produces a toxic contaminant that is released into the local environment, then the risk benefit question is more difficult to analyze. Such was the case with the Dow Chemical plant located in Midland, MI, when the chemical known as dioxin was released into the Saginaw Valley.

4. *Radon* creates an interesting issue in which there is a perceived risk with no known attendant benefit. The risk comes from people building homes and other structures in an area where there is a natural leakage of radon gas into the basement. Too much exposure to the resulting radiation can cause an increase in lung cancer. How much is too much and what action should society take, if any, to alleviate the natural exposure to a risk that has no known benefit?

An overriding characteristic of the list is that a large segment of society experiences the risk while only a small group reaps any significant benefit. For example, aerosol cans are a convenience item used by only a few, while the possible risk from a reduced ozone layer will be borne by all members of society. Socioscientific issues are further distinguished by four more characteristics.

1. There is a deep public interest in the resolution.
2. The information and understanding required to come to a rational judgment are complex and difficult to evaluate.
3. The risk is usually of an involuntary nature where the individual or group exposed to the risk has no choice. This should be compared to the risk taken by a tobacco user where the risk is voluntary.
4. Sound final judgment requires fine tuning and a balancing of several different options. These are concerns about which people may have widely varying attitudes and feelings.

The last characteristic dealing with quality-of-life decisions is the most difficult to handle. How these questions are analyzed and decisions made has given rise to a new procedure called risk benefit analysis. The famous essay "The Tragedy of the Commons" (Hardin, 1968) illustrates the problem. In this article Hardin concluded that the dilemma has no technical solution, but requires a fundamental extension or change of morality. The essay deals with a pasture that was open to all herders. Each herder would be expected to keep as many animals as possible on the commons. This type of arrangement works well since wars, poaching, and disease keep the numbers of both animals and humans well below the capacity of the land to support them. However, the day of triumph finally arrives when the long-awaited goal of social stability becomes a reality. At this point, as Hardin states, "the inherent logic of the commons remorselessly generates tragedy." As a rational person, each herder seeks to maximize his or her gain and concludes that the only sensible action is to add more animals. However, this is the same conclusion that all herders reach. The tragedy is that the system compels each person to increase their herd without limit in a pasture that is limited. Belief in the freedom of the commons brings ruin to all. Hardin then goes on to discuss other commons such as air and water, and the problems society has with pollution. Rational people (similar to the herders) find that their share of discharging waste into the commons is a charge borne by all people. Since this is true for everyone, society is again locked into a system that is destined to bring ruin. One solution to these problems has been to establish laws and regulations that in effect put fences around the commons. Through such mutual coercion, society seeks to break out of the inevitable consequences of pursuing a "common-type philosophy." The resolution of the risk benefit question depends on having rules and regulations in place that protect society as a whole. There are many examples where such coercion is beneficial. For example, how much air pollution is acceptable to enjoy the benefits of greater automobile travel? The debate between the automobile industry and government regulators regarding the fleet mileage is a similar situation.[1] The companies would like to build bigger cars (more profit per unit); however, this would increase the fleet mileage beyond the regulated level and cause greater air pollution due to the larger engines. Without government control there is no question about which way the auto companies would go. Mutual coercion becomes a necessary adjunct in seeking the maximum benefit with an acceptable level of risk for society.

How society presently solves the risk benefit equation will be discussed under the following headings.

Historical
Solving the Risk Benefit Equation
   The Legal Approach
      Criminal
      Civil
      Legislative Process
   The Scientific Approach
   The Rules of Reason
Conclusion

---

[1] Fleet mileage is the miles per gallon allowed for the automobile company.

HISTORICAL

To better identify some important characteristics of risk benefit, the issue needs to be examined from the historical perspective. World War II may be regarded as the first time socioscientific questions were addressed seriously. Before that period there was less need for such analysis, since the time between identifying and reducing to practice any new technology was measured in decades. This gave people a chance to consider all the options before being exposed to the new practice. In addition, the population density was low so that individual or group activities had minimal impact on their neighbors. For example, there was sufficient distance between factories so that streams could refresh themselves of any pollutants before downstream people used the water.

World War II irrevocably changed this pattern. By channeling resources and research into the war effort, the industrial nations achieved enormous advances in every scientific field. What is equally important is that the time between discovery and utilization was shortened. Wessel (1980) identified the war as the first period in the development of the risk benefit concept. The symbol of this new age became the mushroom cloud created by dropping the atomic bomb on August 6, 1945.[2]

The First Period (1945–1960)

This era is characterized by optimism. With the discovery of atomic power, a new age was entered where everything seemed possible, since unlimited power from all the new technology would be available to everyone. Another example may be represented with the synthesis and introduction of DDT, the "miracle bug killer." Farmers believed that they could now raise plants and other food materials free of insects. With this new belief, society felt that by controlling the direction of scientific endeavor, any predetermined future could be achieved. While other nations worried about nuclear holocaust, the response in the U.S. was "We have the secrets, everyone else is ignorant and we will be the protectors of the world." The easy answer to security was SECRECY. This reasoning was partly responsible for the infamous decade of McCarthyism and the witch hunt for suspected Communist spies. Few people at the time appreciated the fact that the U.S. had no permanent monopoly on scientific information.

---

[2] Readers who were born in the 1920s or 1930s can probably make their own identification with this period. My substantiation comes from my farm experience. Prior to 1939, horses were used as a source of pulling power. During the war, most young men were in the service and therefore it became necessary to produce more food with fewer people. The solution was the introduction of tractors. Since farm implements had not yet been developed for tractors, the tongues of the previous horse-drawn equipment were simply shortened with a saw. This act may be regarded as the watershed between the slow-paced technology and the ever-changing technology of the present.

The Pessimistic Period (1960–1969)

Several factors came into focus to initiate the pessimistic period. First, by the end of the 1950s other nations had the secrets of the atom, e.g., on August 12, 1953, Russia set off its first atomic bomb. Second, the reliance of the U.S. on fossil fuel had grown exponentially. Third, the chemical industry had found ways to turn some of these hydrocarbons into useful pesticides, so that agricultural production jumped.

With this activity came the sobering realization that the direction of science could not be controlled by government regulation. Rachel Carson had published *Silent Spring* (Carson, 1962), warning of the coming disaster to the environment with the continuous use of certain chemicals such as chlorinated hydrocarbons. Ralph Nader published *Unsafe at Any Speed* (Nader, 1965), a book highly critical of the manner in which he felt General Motors was ignoring safety for the benefit of profit. These were a few of the reasons that brought on this period of pessimism. Society's easy answer for these troubling questions revolved around the following:

- Ban the use of nuclear power
- Prohibit the use of chemicals
- Etc.

Obviously, there was a clear and growing disenchantment with science. With this disenchantment came a desire to stuff the genie (new technology) back into the bottle.

The Age of Realism (1969 to the present)

The modern period began when Congress reacted and in 1969 passed the National Environmental Policy Act (NEPA). Whatever the original intent, this act became, and still is, a landmark piece of environmental legislation. Among other things it contains a provision mandating an Environmental Impact Statement for any major federal action. Such a statement must describe all risks and all benefits associated with the action before the project is initiated. This is in sharp contrast to the absolute approach of the two earlier periods. As Wessel states, "We have come to recognize that science is neither good nor bad. It is how we use it that counts."

SOLVING THE RISK BENEFIT EQUATION

The risk benefit equation poses one of the most difficult problems that society faces. The right questions can be formulated; however, the questions do not elicit answers, only more questions. The answers vary from person to person, and solutions that satisfy everyone are difficult to find. For this reason government regulations become necessary to help achieve answers that benefit most members of the affected community. Consider, for example, how much should be paid to clean up a hazardous dump site. Clearly the cost will reflect what goals society has in mind. If the dump site will become a future playground for children, the cost of cleanup will be more than if it remained as a storage location for waste materials. Alternatively, the closer people live to the site the greater the perceived risk and the more they would pay than if they lived in the next county. Resolving these conflicting goals and establishing costs are very complex. The problems can quickly become worldwide as opposed to local. The release of chlorofluorocarbons into the stratosphere might alter global weather and climate patterns. Resolution of this type of problem requires that nations learn how to cooperate and take action. For some nations the perceived benefit from continued release of the chlorofluorocarbons may outweigh the long-term risk to the ozone layer. Reaching a consensus on the global scale requires a great deal of patience and perseverance. Consequently, the Montreal accord on the control of chlorofluorocarbons was a landmark achievement (Zurer, 1988). This was one of the first examples where the nations of the world agreed on a common course of action. For further discussions of this accord see the case study dealing with ozone in Chapter 9. Traditionally problems involving conflicts have been solved using the dispute resolution process, described in the next section.

The Legal Approach

Most disputes use some form of an adversary process to reach a solution. The process begins when two parties have a confrontation. Most of these disputes are disposed of by discussion, negotiation, compromise, and settlement. Sometimes the dispute is not resolved. Once positions become hardened, the parties involved resort to lawyers and finally to the courtroom where the accused "has his or her day in court." The adversary process seeks justice and fairness where justice is defined in terms of procedures employed by the lawyers. This provides a series of objective tests to decide if justice was done. For example, did the party have a right to a lawyer, did the judge show prejudice, was there a right to confrontation of witnesses, etc.? When these were met, then justice was done despite the result. One main assumption in this system is that justice, as defined above, is equated to truth. Consequently, truth will be discovered by permitting those who have the greatest stake to fight it out in court using all the "procedures" at their disposal. This is a different definition of truth than that given by the dictionary[3] where truth is in accordance with the accepted facts.

---

[3] Dictionary definition of truth: (1) that which is in accordance with the fact or facts, (2) a fixed or established principle, law, etc.; proven doctrine; verified hypothesis; a basic scientific truth.

There are three approaches to settling legal disputes, each of them serving a different purpose:

1. Resolution of criminal charges against one or more individuals by the State (using State in the sense of any government body).
2. The settlement of disputes of a civil nature between private parties or between a private party and a government body. Civil charges are those charges where an incident has occurred in which no law has been broken, but where damages have been inflicted. An example illustrating the difference between criminal and civil would be an incident in which a person runs a stop sign and crashes into another car. The criminal action is running the stop sign as there are laws relating to such matters. The civil action is recovering the damages that have occurred to the car.
3. Resolution of disputes or conflicts between society and individuals or private groups through the legislative process.

The processes and objectives for dispute resolution in the above areas will be described below.

Criminal Cases

In the field of criminal law, the duty of the court (whether judge or jury) is to decide whether a person or corporation is guilty of committing a crime (i.e., whether a person violated a statute enacted for the protection of society). The American and English criminal law system is based on the premise that no person should be deprived of life or property without due process of law. Due process was adopted to protect citizens against the tyranny of the majority. In other words, it could be that in a given situation every citizen in a community believes that an individual is guilty of a crime and should be punished. Under our system the person must be proven guilty in a court of law, observing due process and regard for that individual's rights. Indisputable evidence of guilt may have been obtained by violating the constitutional right of the individual against unreasonable search and seizure. That evidence may not be used to convict the individual. The criminal law system has adopted a bias that protects the innocent. If there is a chance of a mistake, the benefit of the doubt should remain with the accused and not with the state.

An example will make these arguments clearer. On June 8, 1979, the New York State Supreme Court overturned the first-degree murder conviction of Tony Provenzano. There was no substantive question as to the guilt of this man. The ground for reversal was that Provenzano had been denied procedural rights to which he was entitled. What had happened was that a lower court judge had improperly refused to excuse a potential juror. This juror had met with the prosecutor at a political club. To excuse this juror, Provenzano's lawyer had been forced to use one of the limited number of peremptory juror challenges. As a result he lost a challenge he might later have used against one of the 12 jurors finally selected who voted unanimously to convict. The State Supreme Court ruled that Provenzano had been denied his right to a fair trial and that the decision required reversal. It made no difference at all whether the evidence was sufficient or even overwhelming to convict. The public accepted this ruling with very little comment other than that the judge should have known better. While there may be times as in the above example when the situation appears unreasonable there is no question that the bias is where it should be.[4]

Civil Disputes

Civil disputes in contrast to criminal suits are not about breaking a law, but about collecting damages. Often the latter is more complicated. The two disputes are similar in the sense that neither process was designed to create policy. In the traditional view, even the U.S. Supreme Court should decide individual cases based on applicable law using any precedent that might have a bearing. The Court should not reach decisions on broad policy questions in an advisory way or in a way that would have consequences beyond the immediate case before them.

The following example dealing with a civil dispute shows the weakness in the court system in addressing socioscientific questions. On May 25, 1979, American Airlines Flight 191, a DC-10 bound for Los Angeles, crashed on takeoff from Chicago's O'Hare airport, killing 275 people. The Federal Aviation Administration (FAA) failed to act quickly in grounding the planes and the Airline Passengers Association filed suit asking that the FAA be ordered to ground all DC-10s until the cause of the accident was found. The FAA's attorneys resisted on procedural grounds. The lower court issued an injunction ordering the grounding of the DC-10s. The U.S. Court of Appeals for the District of Columbia decided that they had jurisdiction in the case and overturned the lower court's decision. The FAA was correct so far as procedural rules were concerned. However, they were not correct in understanding how emotionally involved the flying public was in the issue.

The reaction to the civil case and the criminal case as described above differed in the following way. The public was intimately concerned with the DC-10 crash and wanted the situation resolved quickly, while in the case of Provenzano the public was only mildly interested and could live with a guilty or a not guilty decision. These two cases illustrate the fatal defect in the dispute resolution process. The misplaced emphasis is on procedure. What is wrong? The adversarial process seeks justice and fairness, not truth. In

---

[4] It should be pointed out that Provenzano was later tried and convicted on another charge and died in prison.

socioscientific controversies the public wants the focus on the truth, i.e., whether the planes are safe. The distinction is quite important. The public wants a resolution process in which real and vital concerns are properly presented and discussed. So far these wishes are not being met. This leads to the third method of resolving legal disputes.

The Legislative Process

This is the area that deals with the balancing act between society and the individual or private parties. Acid rain is an illustration of this area. For example, owners of a property in upstate New York might bring a civil action against a steel mill in Ohio in the hope of recovering damages for injury to their property. In doing so it would be necessary for them to establish that the Ohio plant had breached a duty that it owed to that individual. The assertion would be that the Ohio plant could reasonably foresee that their action would become the proximate cause of injury to the person in New York. In that event a jury could decide that the plant was liable for damages. This would not, however, solve the problem of emissions from that plant. If the owners wished to continue taking the risk of individual suits, there would be nothing to prohibit them (without legislation) from continuing their operation, that is, the decision in the civil case would not create any policy determinations that the plant should stop emissions. This would have to be done in the legislative arena.

The primary means of resolving many of these socioscientific issues is through the legislative process and not through the court. These laws then are administered through regulations made by public officials. The courts are intended to deal with specific incidents only and should not be viewed as an important part of the general balance between public and private interest. Courts primarily follow rather than make the laws.

The major problem with the legislative approach is the introduction of more rules and regulations. By controlling our lives through these rules, we simultaneously give up a certain amount of freedom. Maximizing both freedom and security is impossible. If there is a maximum amount of freedom, there will be minimum security, and vice versa. As society reluctantly accepts more government regulation and control there must be a recognition of the loss of freedom that results. The goal must be to achieve the minimum amount of risk with the maximum amount of freedom. When this does not happen voluntarily, then government regulations must be accepted.

Scientific Approach

The adversary process has become so pervasive that people have forgotten that there are other ways to solve problems. One way is with the scientific method. At one time this method was very objective and devoid of emotion. Today investigators are so tainted with a desire to succeed that the phrase "adversary science" is beginning to creep into our vocabulary.[5]

The *New Columbia Encyclopedia* describes the scientific method as follows:

> Information or data is gathered by careful observation of the phenomenon being studied. Based on that information a preliminary generalization or hypothesis is formed, usually by inductive reasoning, and this in turn leads by deductive logic to several implications that may be tested by further observation and experiments. All of the activities of the scientific method are characterized by a scientific attitude, which stresses rational impartiality. Measurement plays an important role, and when possible the scientist attempts to test his theories by carefully designed and controlled experiments that will yield quantitative rather than qualitative results. There is no place in this process for personal attack. The scientific method requires the open dissemination of results, where there is an effort to achieve scientific consensus.

The differences between the scientific and the legal approach are striking, as the following chart shows:

	Scientific	Legal
Method:	Nonadversarial, peaceful	Adversarial war
Objective:	Discovery; consensus "truth"	Win
Technique:	Analysis, testing validation replication, disclosure	Persuasion "tactics"

Once science has accomplished its goal, the problem emerges of communicating the results in an understandable manner to the lay public. This has been a hurdle to overcome ever since the first scientist made a discovery. For example, convincing the populace that the earth was round took several generations before there was general acceptance. Today the problem is the same but the issues are more intimately connected with lifestyle, i.e., how much wilderness needs to be preserved at the expense of oil exploration, or should there be unlimited access to high technical medical operations regardless of ability to pay? The resolution of these and similar problems does not depend on more science, but in bringing credibility to science. We need "responsibility" in science, just as we need responsibility in the practice of law. Scientists need to become involved with "values" and with "quality of life" and "public

---

[5] This is well illustrated with the incident involving Dr. Baltimore, a Nobel laureate (*Time*, p. 66, May 20, 1991). This event involved the falsification and publishing of an article by Baltimore and one of his associates. While Baltimore defended his study, it turned out that the data had been wrong. The ramifications of this affair will be felt for many years in the scientific establishment.

policy" issues. Unfortunately, they must become adversarial in their approach. What should they do? Many respond by crawling into their ivory tower and refusing to become further involved. This obviously is not the correct approach.

Science and scientists should not withdraw from the public struggle. What they need to do is separate and clearly distinguish between "science" and "value." They must communicate their science to the public and participate as a full partner in the decision process.

The line between communicating science and communicating value is difficult to define conceptually. One illustration is the story of the heart-disease patient who obtains two opinions from two well-known heart surgeons. Both doctors have the same information, but the message communicated to the patient is based on the value system of the individual doctor (Wessel, 1980). One doctor believes that the risk of surgery is insufficient to prevent the patient from having the same active lifestyle as before and recommends open heart surgery. The second doctor believes that the restriction on one's lifestyle because of the ailment is not worth the risk of an operation and recommends a change of lifestyle to accommodate the disease. What is lacking in these two recommendations is an explanation of the facts, including the risks, so that the patient can make his own decision. Such an approach would be very time-consuming for a busy doctor, and is usually ignored. Another example is illustrated by toxicologists saying, "The risk is acceptable." They are stating a conclusion that includes toxicological data, research, and opinion about which they are uniquely qualified to comment, and a "value" or "public policy" conviction about which they are no more qualified to make a decision than the lay person. These two issues must not be confused. When they are, it appears that the professional has some special right to decide "quality of life" matters for others. Besides separating value from factual conclusions, scientists and other professionals must make a valid effort to express the uncertainty in their findings to the people who are not familiar with their discipline. There are many examples of this conflict.

1. Pathologists have a difficult time agreeing on whether a tissue is benign or not. They must discuss their results, including the reasons for their uncertainty.
2. Analytical chemists will argue about the significance of the signals on their machines. Is it random noise or do the signals signify a chemical?

Such phrases as, "There is substantial risk of malignancy," or "There is no evidence of serious contamination," are conclusions based on the speaker's value system. The pathologist must explain that "substantial" means "more than 10% chance" (assuming that to be the recognized test) and that for the analytical chemist "no evidence" means "below a signal-to-noise ratio of two to one." The professionals in these respective fields know what these phrases mean but the lay public does not. Increasing the understanding of science by the lay public will require a major effort. A beginning may be made by having scientists translate their findings into nontechnical language.

Rules of Reason

The major difference between the legal and the scientific approach to dispute resolution is that the scientist must consider all available information while the lawyers use only the arguments favorable to their client. This difference makes the scientist appear indecisive while the lawyer is uncompromisingly clear. Neither approach is suited to solving the socioscientific problems similar to the DC-10 crash. The flying public is not interested in the adversarial clash between opposing lawyers who are using all the weapons at their command to defend their respective clients. Similarly, they are not interested in the scientific discussion of all possible reasons for the crash. What they want is a discussion focused on the issue, "Is it safe to fly on a DC-10?" What is required is a combination of the two techniques. Wessel has proposed a method that attempts to accomplish this objective. The proposal, called "The Rules of Reason" (Table 1), requires that improper methods be avoided and that all parties act responsibly. There are many examples to illustrate the tactics that need to be avoided.

1. When General Motors and Nader had their dispute over the Corvair, GM attacked Nader personally rather than face the issue of the Corvair. While this is a legitimate procedural approach, i.e., discredit the witness, the approach backfired because GM lost all credibility. Such an attack does not serve the public. The fact that GM should have used was that in 1967 the Corvair had the lowest single car accident rate of 0.16/million miles compared to 0.24 / million for other cars. Based on a value system, GM was poorly served by its attorneys who relied on "courtroom tricks."
2. When the Ford Motor Company had its problems with the Pinto, it became a classic case of what not to do. This was a situation similar to the Corvair, and Ford followed the old rules of the adversarial process. The company "lost" documents, dragged its feet on procedural matters, and used a variety of ploys in the hope that the plaintiff would tire of the process and go away. It would have been better for Ford and the driving public to have discussed and resolved the issue of the safety of the Pinto.

In a similar manner, CBS is using the same tactics in its lawsuit with the apple growers (see the case study on Alar in Chapter 9). The media giant is dragging the issue through lengthy hearings in an effort to wear out the

Table 1. The Rules of Reason

Data will not be withheld because "negative" or "unhelpful"

Concealment will not be practiced for concealment's sake

Delay will not be employed as a tactic to avoid an undesired result

Unfair "tricks" designed to mislead will not be employed to win a struggle

Borderline ethical disingenuity will not be practiced; motivation of adversaries will not be unnecessarily or lightly be impugned

An opponent's personal habits and characteristics will not be questioned unless relevant

Whenever possible, opportunity will be left for an opponent's orderly retreat and "exit" with honor

Extremism may be countered forcefully and with emotionalism where justified, but will not be fought or matched with extremism

Dogmatism will be avoided

Complex ideas will be simplified as much as possible to achieve maximum communication and lay understanding

Effort will be made to identify and isolate subjective considerations involved in reaching technical conclusion

Relevant data will be disclosed when ready for analysis and peer review—even to an extremist opposition and without legal obligation

Socially desirable professional disclosure will not be postponed for tactical advantage

Hypothesis, uncertainty, and inadequate knowledge will be stated affirmatively—not conceded only reluctantly or under pressure

Unjustified assumptions and off-the-cuff comment will be avoided

Interest in an outcome, relationship to a proponent, and bias, prejudice, and proclivity of any kind will be disclosed voluntarily and as a matter of course

Research and investigation will be conducted appropriate to the problem involved; although the precise extent of that effort will vary with the nature of the issues, it will be consistent with stated overall responsibility to solution of the problem

Integrity will always be given first priority.

From *Science and Conscience* by M.R. Wessel, Columbia University Press, New York, 1980, p. 97. With permission.

plaintiffs. The public would be better served if the trial focused on the main issue.[6]

3. When IBM settled its antitrust suit with Control Data, one condition of the settlement was that Control Data destroy a computerized database that it had used very effectively in its lawsuit against IBM. This was done much to the unhappiness of several other competitors who were relying on the presence of the data in their own lawsuits against IBM. While IBM was within its legal rights to have this evidence destroyed, the public was upset and IBM lost considerable credibility.

The approach to solving these and other issues is intimately tied to the rules that are reproduced in Table 1. If these rules are adopted, there is no question that the public will be better served in the resolution of the many complicated socioscientific problems that face society today.

CONCLUSION

It is becoming increasingly clear that society desperately needs a process for resolving complex socioscientific issues. Three forms of problem solving have been presented: the legal approach, the scientific approach, and a combination of the two known as the rule of reason. It has been argued that the first system needs altering to deal with catastrophes such as the DC-10 crash. The problem with the second system (the scientific approach) is that it does not stay focused on the main issue of concern to the public, i.e., the scientist becomes immersed in the various reasons for the failure rather than answering the main question, "Is the DC-10 safe to fly?" One approach is to combine the best of the legal and scientific methods. Wessel's Rule of Reason is an attempt at such an amalgamation that has the possibility of dealing with these difficult problems. The alternative would be to rely more on government regulations. For those who value freedom, such reliance becomes very distasteful. The choice needs to be made before it is made for us.

The four case studies to be presented in the concluding chapter have been selected to illustrate how the principles of environmental risk assessment apply to real situations.

---

[6] It would appear that the strategy is working in that the U.S. District Court in Washington State has dismissed the lawsuit. At the present time (October 1993) a decision to appeal has been made.

9   CASE STUDIES

Four case studies have been selected to illustrate the principles developed in the previous chapters. In addition, the topics have been chosen to portray different scenarios.

Dioxin—This chemical was a contaminant in the defoliant known as Agent Orange used extensively during the Vietnam War. Many veterans claimed health problems from dioxin exposure that occurred during the application of the herbicide to the jungle. With the widespread publicity resulting from the various lawsuits, dioxin became a national issue. Rather than dealing with the entire complicated political and scientific problem (Gough, 1986), the present study focuses on a small issue. The case involves the distribution of dioxin in the watershed where a major producer of products containing dioxin is located, and assessing the risk to the inhabitants.

Radon—The second case study involves the issue of radon gas contaminating homes in various parts of the country. This is an interesting study because the risk is natural in origin and has no economic benefit.

Ozone—The destruction of the stratospheric ozone layer is a risk shared by all people of the world. The attendant benefit from the continued use of aerosol propellants and refrigerants, on the other hand, is shared by only a small percentage of the population. Another interesting feature of this case is that the chemicals by themselves are very innocuous. However, the impact that they have on the stratospheric ozone could have serious consequences.

Alar[1]—The final case to be presented is a situation where apples had been exposed to an alleged chemical carcinogen. On the benefit side, the consumers had a more economical and aesthetically pleasing product. How these two issues, chemical contamination in food and the economics of production, are balanced is a very important issue not only for Alar, but for other chemicals that might be found in the food supply.

Each case study will be discussed under the following headings.

---

[1] Alar is a registered trademark of the UniRoyal Chemical Co., for the product containing the chemical daminozide.

Historical Background

The events leading to the situation where the public became aware of the problem and demanded action will be described.

Risk Assessment

This will follow the guidelines established by the National Academy of Sciences (NAS, 1983) and endorsed by the U.S. Environmental Protection Agency (EPA, 1986). The flow diagram introduced as Figure 1 in Chapter 1 will help clarify the procedure to be followed in each study.

Hazard Identification

This is the process of determining if the chemical or agent has the necessary properties to cause an increase in the incidence of a health condition and/or any environmental impact other than health. In order to accomplish this task, the physical and chemical properties are assessed along with the toxicological effects. To complete the identification, possible routes of exposure need to be examined. The results of this analysis will show the type of hazard that might exist from being exposed to the chemical.

Dose Response Assessment

Following the hazard identification, the relation between dose and the incidence of the effect needs to be studied. Dose response investigations usually require extrapolation from a high dose to a low dose and extrapolation from animals to humans.

Exposure Assessment

What are the sources of the chemical, and what exposure can be expected for the population at risk?

Risk Characterization

Finally, the exposure is matched with the dose to establish the size of the potential risk.

Conclusions

What has been learned and can it be applied to similar cases as they emerge in the future?

CASE STUDY 1: DIOXIN

Historical Background

The chemical known as tetrachloro dibenzo-p-dioxin (TCDD) has been characterized as the most toxic chemical ever made. The fear and anxiety created by the possible exposure to this material had its origin in the Vietnam War. During the war, the U.S. Department of Defense sprayed a herbicide called Agent Orange on the foliage of the trees. The purpose of the spray was to defoliate the jungle, making it more difficult for enemy snipers to remain concealed. In the late 1960s the discovery was made that Agent Orange was contaminated with a material called 2,3,7,8-Tetrachloro Dibenzo-p-Dioxin (Structure I). The chemical turned out to be very toxic to laboratory animals, resulting in a great deal of public attention. Since the news media had a difficult time with the correct name, it has been popularly called TCDD or dioxin. In this chapter, dioxin will refer to the series of chemicals related to Structure I, while the name TCDD will be reserved for the specific compound. Many veterans during their active service in Vietnam were accidentally sprayed with the defoliant, (Hansen, 1987). After the war different ailments, including cancer, were blamed on contact with TCDD. This resulted in several lawsuits against the manufacturers of the herbicide.

One of the major producers was the Dow Chemical Company plant in Midland, MI (Figure 1). The waste stream from the manufacturing site entered the Tittabawassee River with the result that dioxin including TCDD has been identified in both the river and the resident aquatic organisms. The case to be examined will deal with assessing the risk to people living in this watershed.

In hindsight, the first indication that there was some toxicity associated with certain chlorinated products was in the late 1950s, when the death of

2,3,7,8 - Tetrachloro Dibenzo -p- Dioxin (TCDD)
I

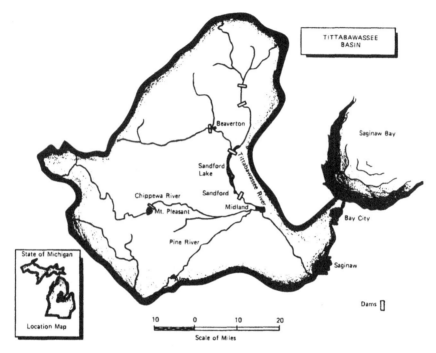

Figure 1.   Map of the Tittabawassee watershed in Michigan.

several chickens was reported. At the time the demise of the exposed poultry was unexplained. Subsequent investigations have led scientists to the conclusion that the loss was a result of feed contaminated with a wood preservative containing pentachlorophenol. For many years the nature of the contaminant was not known. However, more recent studies have identified the impurity as hexachloro dibenzo-p-dioxin (Structure I with six chlorines on the aromatic ring). By 1969 there was concern over the adulteration of other chlorinated products. Links were quickly made to 2,4,5-trichlorophenol and its use as an intermediate for the manufacture of both the herbicide and the antimicrobial agent, hexachlorophene. Further evidence showing a problem was the observation that workers involved in the production of these materials developed a skin rash called chloracne. The same rash had been reported by workers in the production of PCBs (polychlorinated biphenyls).

The fact that the highly toxic contaminant TCDD was present in significant amounts (μg/g or ppm) combined with the observation that the phenoxy herbicides had been widely used in the U.S. on rangelands, rice paddies, and forests prompted further public concern and debate. The debate over the use of these herbicides continues. Only four of the many incidents will be mentioned.

1.   A group of citizens conducted their own survey of miscarriages in Oregon in 1979 that implicated 2,4,5-T from forest spraying.

The survey was later discounted through a more detailed scientific study. However, the damage had been done as the possible connection of miscarriage and 2,4,5-T had been made. This is an example of the heuristic of representativeness (Chapter 1). The people in Oregon were convinced that there was a positive correlation between exposure to the chemical and to the reported miscarriages and no scientific study could convince them otherwise. The result of this incident in combination with other reports led the EPA to ban the use of 2,4,5-T for vegetation control. The EPA's action was noted in Canada, where the product was subsequently banned by the various provinces. Only in the province of Nova Scotia did the court find no evidence to prove the alleged risks and allowed spraying to continue.

2.   The second incident was a major accident in Seveso, Italy, in July 1976. This occurred in a plant that used trichlorophenol as one of the products. A reaction vessel exploded and caused a widespread distribution of chlorinated phenols including many chlorinated dioxins. Because of this accident, many children developed a skin rash known as chloracne. In addition numerous animals in the immediate area died. An extensive health surveillance system was put into effect to record any continuing and long-term effects of dioxin exposure in the population. Currently no cases of cancer related to the incident have been observed (Letts, 1991).

3.   The third story relates to the concerns of the people living in Times Beach, MO. Still bottoms containing wastes from the production of 2,4,5-trichlorophenol and PCBs were used in 1982 as a spray to contain the dust on several roads in and near the town of Times Beach. Besides the roads, the residues were also sprayed in a horse arena for the same purpose. Because of this activity, several horses died and children that had played in the arena became sick. During the winter of 1982 severe floods took place that caused the spread of the dioxin-contaminated soil through the residential area of the town. The continuing controversy caused the government to buy the town and relocate the 2232 citizens at a cost of $33 million. An editorial in *The Wall Street Journal* reported that Dr. Houk, director of the Center for Environmental Health and Injury Control, now believes that the evacuation of the town was unnecessary.[2] The full extent of the problem is still being debated.

4.   The final incident that will form the basis for this case study was the discovery of dioxin in the Tittabawassee watershed (Figure 1). Dow Chemical operates a major manufacturing plant located on the river and, since the company was a major supplier of the

---

[2] *The Wall Street Journal,* Aug. 6, 1991.

defoliant, the two terms dioxin and Dow became closely associated. With all of the adverse publicity it is no surprise that the announcement of TCDD being found in fish in the Tittabawassee River made the evening news. The analytical studies performed by investigators at Michigan State University were released in March 1983 (Kaczmar et al., 1983). Shortly after the first news release, Region 5 of the EPA (located in Chicago) issued a report suggesting that the Dow plant located in Midland was the largest single source of dioxin in the Lake Huron watershed (EPA, 1983).

Over the next six months there was an average of two or three articles a week on the "Dioxin" story. Besides the articles there were editorials and cartoons associating Dow with dioxin. The next section will present the data to help formulate the possible risk that people face from being exposed to dioxin in the area shown in Figure 1. How bad is the contamination, and is the local population subject to an adverse risk? These and other questions will be addressed in the remainder of this study.

Risk Assessment

Following the outline in Chapter 1, the analysis of risk assessment will be dealt with under the headings listed below:

Hazard Identification
Dose Response
Exposure Assessment
Risk Characterization

Hazard Identification

There are three major topics that need to be addressed in identifying the potential hazard of a chemical. As mentioned earlier, these are: physical and chemical properties, toxicological effects, and route of exposure.

Physical and Chemical Properties — The confusion over the single name of dioxin representing a multitude of isomers has created a situation where great care needs to be taken in reading the articles on this subject. The care is needed because the toxicity of the various homologs varies by orders of magnitude (see the following section dealing with toxicology).

The main isomer, and the one that has the greatest toxicity, is a trace contaminant that is inadvertently produced when 2,4,5-T is prepared from

Table 1. Chemical and Physical Properties of TCDD

Molecular weight	322
Solubility	0.2 μg/L
Vapor pressure	9.3×10^{-11} kPa
Partition coefficient (octanol and water)	~1,000,000
BCF in trout	51,600[a]

[a]Value used by EPA in deciding the water quality criteria number. The high value of 51,600 has been questioned by many investigators and there is no consensus which of the many values is correct. At present numbers, ranging from a low of 1875 to a high of 51,600, have been found. The high value is a very conservative approach and will be used in the present analysis.

2,4,5-trichlorophenol, as the following reaction shows:

2, 4, 5-Trichlorophenol 2, 4, 5-Trichloro phenoxy acetic acid

TCDD

Besides this molecule there are several other isomers, most of which have been identified and studied. This chapter will only discuss the properties for TCDD summarized in Table 1.

What has made this issue very complex is that for many years the environmental residues of TCDD were difficult to detect. In the early 1970s, detection limits for toxic substances were usually in the μg/g (ppm) range. Before that time, toxicologists would take wipe tests[3] in the manufacturing plant and use an animal bioassay to detect the presence of the contaminant. The response

---

[3] These were tests where the toxicologists would wipe the top of the bench or walls with a piece of cloth. The cloth then would be used to expose animals to the contaminated material. If the chemical was present in high enough dose it would cause acne-like blisters to appear on the skin of the animal and the analyst would report a positive finding. Obviously, very little could be said about the concentration.

Table 2.   Acute Toxicity for TCDD in Various Animals

Animal	μg/kg
Guinea pig	1
Mouse	114
Rat (males)	22
Rat (female)	45
Monkey	70
Hamster	5000
Dog	>300

Note: Other chlorinated dioxins are much less toxic, e.g., 2,3,7-trichlorodioxin is 20,000 times less toxic than TCDD.

they were looking for was acne, a skin disorder similar to juvenile acne, where open sores would develop. However, they had no idea about either structure or concentration of the active agent.

One fallout of the controversy surrounding TCDD was the development of sophisticated analytical methods for detecting these molecules at lower and lower levels. Initially, the methods were based on the above bioassay and then procedures were developed to detect at the μg/g (ppm) level.[4] Now they have moved into the pn/g (ppb), pg/g (ppt), and fg/g (ppq), six and nine orders of increased sensitivity and still improving (Lamparski et al., 1979). The quest for greater sensitivity for TCDD has had a major influence on analytical methodology for these and other trace contaminants.

Toxicology — The comparative single oral dose LD_{50} for TCDD is shown in Table 2 (summarized by Letts, 1991). As shown, the acute lethality for guinea pigs is very low and is the reason that TCDD has been called the most toxic chemical ever made. Tschirley concluded from data on inadvertent exposure to humans that man is less sensitive to TCDD than certain laboratory animals, such as the guinea pig (Tschirley, 1986).

However, it is the carcinogenicity of the molecule that is of the most concern. Table 3 summarizes the 2-year study (Kociba et al., 1978). This is still the most definitive investigation and provides the data that has been used by the regulators to estimate the risk based on the procedures discussed in Chapter 4.

Route of Exposure — There are many routes of exposure for dioxin. The troops in Vietnam were exposed largely through the skin while the children playing in the dirt at Times Beach, MO, were exposed by ingesting the

---

[4] Recall the $\mu g = 1 \times 10^{-6}$ g, $pn = 1 \times 10^{-9}$ g, $pg = 1 \times 10^{-12}$ g, and $fg = 1 \times 10^{-15}$ g.

Table 3.   Summary of TCDD Carcinogenic Data (Kociba et al., 1978)

μg/kg/day	Response	$R_c{}^a$
Control	9/86 (0.1087)	
0.001	3/50 (0.06)	
0.01	18/50 (0.36)	0.28
0.1	34/48 (0.71)	0.67
0.001 μg/kg = 1 nanogram = 1×10^{-9} grams		

[a]This is the response corrected for the number of tumors appearing in the control group (see Chapter 4 for more details).

contaminated soil. The route of exposure that will be of interest in this study will be oral. It is assumed that individuals living in the Tittabawassee watershed become exposed by either drinking the water, eating the fish from the river, or a combination of the two. Using these routes of exposure, a model has been developed to assess the risk from contact with dioxin.

Dose Response

The data in Table 3 establish a dose response for TCDD. Recall from Chapter 4 that the first step in analyzing such data is to make a correction for the background tumor incidence shown by the control (i.e., 9 out of 86 animals in the control group had tumors). Using Abbott's formula the correction is given by Equation 1.

$$R_c = \frac{R - control}{1 - control} \tag{1}$$

The corrected responses are shown in Table 3 and graphed in Figure 2. For the lower doses (less than 0.001 μg/kg/day) there are no experimental data. The problem that presents itself is to take the observed response at 0.01 and 0.1 μg/kg/day and extrapolate, usually by several orders of magnitude, to the selected risk level.

The conservative approach assumes that TCDD is a cancer initiator. Consequently, there must be a linear relation between the lowest dose showing a response and zero (Chapter 4). Thus, the slope of the line in Figure 3 is the risk divided by the dose or 0.28/0.01 for a rat potency factor of 28 (μg/kg/day)$^{-1}$. Since humans are larger than rats, the above factor needs to be corrected for this difference. The conversion normally used is to take the cube root of the ratio of the average body weights of humans to rats (see Chapter 4 for

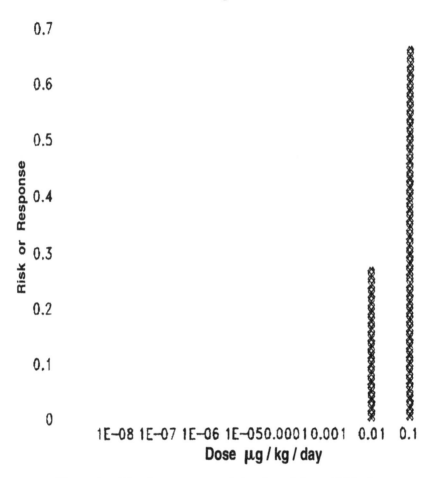

Figure 2.   The dose response data plotted from Table 3.

more details). Thus, the potency for humans is

$$Q^* = 28 \ (\mu g/kg/day)^{-1} \times (70/0.4)^{1/3}$$
$$= 157 \ (\mu g/kg/day)^{-1}$$

If the acceptable level of risk is 1/100,000, then the RSD (risk-specific dose) is the risk/Q^* or $1 \times 10^{-5}/157$ for a value of 6.37×10^{-8} $\mu g/kg/day$ or 64 fg/kg/day.

The literal interpretation is that for a population of 100,000 exposed to 6.4×10^{-8} $\mu g/kg/day$, the upper limit to the increase in number of cancer cases is 1 in 70 years. For a population of 100,000, the number of deaths expected from cancer is 25,000. The exposure to the RSD might increase the expected deaths to 25,001.

A different approach to establishing an RSD is to assume that TCDD is a promoter of cancer rather than an initiator (see Chapter 4). Here, by maintaining the exposure below the experimentally derived no observable effect level (NOEL), a safe level can be estimated. This is done by taking 1/100 of the NOEL, which from the data in Table 3 is equivalent to 1/100 of 0.001 µg/kg/day. The factor 1/100 is designed as a safety factor in favor of humans over rats. The RSD by this technique is 1×10^{-5} µg/kg/day. Consequently, by maintaining the dose at or below the RSD it can be concluded that TCDD will not produce cancer. Note that risk is not mentioned. It is simply stated that below the RSD the exposed population is safe.

Exposure Assessment

There is no doubt that the chemical plant in Midland, MI was responsible for most of the observed contamination seen in the Tittabawassee and Saginaw rivers. The next question concerns the sources of dioxin and the exposure concentrations that result from the various inputs.

Sources
1. Waste streams from the plants producing the various chlorinated products of the Dow Chemical Company:

   - Pentachlorophenol and tetrachlorophenol—these two preservatives are contaminated with hexa-, hepta-, and octachlorodibenzo-p-dioxins at the ppm level.
   - 2,4,5-Trichlorophenol—used as a starting material for 2,4,5-T; contains traces of TCDD.
   - 2,4,5-T—contains TCDD as a contaminant.
   - 2,4-D—contains various dioxins but no TCDD.

   As of 1987 these materials were no longer being made in the Midland plant.
2. Leaching can occur from the waste dumps associated with Dow's production plants. The present operation of the Dow facility guards against the leaching of waste. However, at one time there was considerable leakage of various chemicals including dioxin into the river.
3. Combustion in the incinerator located within the Dow complex. Most of the dioxins produced in the combustion process are the higher chlorinated dioxins (less than 1% TCDD).

The Exposure — The single greatest route of exposure for humans to TCDD in the Tittabawassee and Saginaw watershed has to be the river. The monitoring

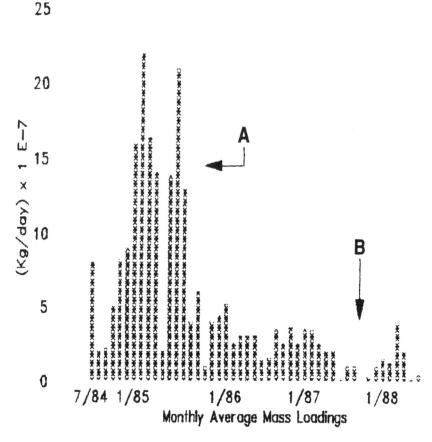

Figure 3. Monitoring data from the outfall at the Dow plant in Midland, MI. A indicates where filtration technology was applied and B where incineration was installed.

data from Dow's major outfall to the river are shown in Figure 3. In addition, the figure also shows the impact that the installations of various devices have had in lowering the amount of TCDD entering the river. Data supplied by Dow (Croyle, 1989) allows for the following calculations.

1. The waste stream has an average flow of 20 million gal/day or 75,600 m³/day. It must be remembered that in all these calculations the importance of checking units needs to be stressed.[5]

---

[5] Do not be caught in the embarrassing position of mixing metric and English units or in using different volume sizes such as cubic meters or liters. The important point is to pick a system and be consistent.

2. The river flow is about 1200 ft³/sec, which converts to 2,930,000 m³/day.
3. The average mass of TCDD discharged in 1985 (Figure 3) was about 15×10^{-4} g/day. These were the years when the greatest public reaction occurred. Since then the Michigan Division of Dow has added several new control strategies that have greatly reduced the mass loading of dioxin to the river. This is seen in Figure 3.
4. Dividing the mass by the volume ($15 \times 10^{-4}/7.56 \times 10^7$) yields a value of 20 pg/L or 20×10^{-6} µg/L for the concentration of dioxin in the discharge.
5. The Michigan Department of Natural Resources (MDNR) allows three fourths of the river flow for dilution. Thus, the dilution factor in going from the discharge to the river is

    (discharge + 3/4 of the river flow)/discharge = 30

    In other words, the stream concentration will be the concentration in the discharge/30 or 0.66 pg/L (0.66×10^{-6} µg/L).
6. The MDNR has allowed Dow to discharge up to 8 pg/L in the waste stream.
7. At a level of 20 pg/L in 1985 Dow was outside the permitted level. By 1988 with the new controls in place the level of dioxin had fallen to about 3 pg/L, well within the state guidelines.
8. One final observation is that the detection level for dioxin in water is 1 to 2 pg/L.

The conclusion from the above is that in 1985 the exposure level in the river was about 0.7 pg/L. The next section will determine the magnitude of the risk associated with such an exposure.

Risk Characterization

Since the river is the largest source of dioxin, it becomes the route by which people living in the valley must be exposed. EPA has established the following model for estimating the exposure from such a route:

An average person weighing 70 kg eats 6.5 g of fish a day and drinks 2 L of water a day from the river for a lifetime of 70 years.

With this model the water quality criterion (WQC) for the river is estimated by means of Equation 2.

$$WQC = \frac{daily\ amount\ deemed\ acceptable}{(amount\ of\ drinking\ water) + (amount\ of\ fish)} \tag{2}$$

Each element of Equation 2 will now be described.

1. The daily amount that is deemed acceptable is simply the RSD estimated previously, multiplied by the average weight of an adult person (70 kg). To estimate a concentration, the mass of TCDD, from the above calculation, must be divided by the volume of water.
2. From the model it is assumed that an individual drinks 2 L of water a day from the river. This is the first component of the water volume.
3. The second component is derived from eating fish that are at equilibrium with the TCDD in the river. By using the bioconcetration factor (BCF), it is possible to decide the water equivalent of the fish consumed in terms of liters per day. In other words, how much water would a person have to drink to equal the amount of chemical present in the fish? The answer is deduced from Equation 3.

$$[Fish\ consumed\ (kg/day) \times BCF\ (liters\ of\ water/kg\ fish)] \tag{3}$$
$$0.0065 \times 51,600 = 335.4\ L/day$$

Thus the WQC is given by Equation 4.

$$WQC = \frac{RSD \times 70\ kg\ person}{(2\ L + 335.4\ L)} \tag{4}$$

The WQC for TCDD acting either as a promoter or as an initiator will now be examined (for a review of these principles, see Chapter 4).

Promoter — Scientists from many countries consider TCDD acts as a promoter. Thus, a threshold exists below which the dose is considered safe. As discussed previously, the RSD is 1×10^{-5} μg/kg/day. This is now multiplied by 70 kg (the average weight of an individual) to arrive at a mass of TCDD consumed in 1 day. The corresponding WQC is derived by dividing the mass of TCDD consumed by the volume of water, according to Equation 4.

$$WQC = 1 \times 10^{-5} \times 70 / 335.4 = 2 \times 10^{-6}\ μg/L = 2\ pg/L$$

Consequently, if the concentration is at 2 pg/L then a person drinking 2 L of water per day and eating about 1/4 pound of fish per week for a lifetime is considered safe from developing cancer via this exposure of TCDD. During 1985, the period of highest discharge, the water concentration was estimated at 0.7 pg/L, well below the estimated WQC for a promoter. It needs to be pointed out that this latter number is only an estimate. The value of 0.7 pg/L is below the analytical sensitivity of 1 to 2 pg/L and impossible to measure at the present time.

Initiator — Most U.S. governmental agencies accept as a policy that there is no safe dose and that all carcinogens act as initiators. Accordingly, the MDNR established 1/100,000 as the appropriate risk level. The calculation in the previous section showed that the RSD for an initiator was 6.4×10^{-8} µg/kg/day. Similar to the above calculation, the WQC number now becomes

$$WQC = 6.4 \times 10^{-8} \times 70/335.4 = 1.3 \times 10^{-8} \text{ µg/L} = 0.013 \text{ pg/L}$$

During the 1985 period, the level of dioxin in the river was about 50 times this level. In other words, using the assumption that TCDD is an initiator, the calculated risk exceeded the allowed level of 1/100,000. The actual risk in 1985 for a water concentration of 0.66×10^{-6} µg/L may be estimated as follows:

1. From Equation 4 the dose for a 70-kg person is 0.66×10^{-6} µg/L $\times$ 337.4 L/day/70-kg person or 3.2×10^{-6} µg/kg/day.
2. Multiply this dose by Q*, the cancer potential of 157 (µg/kg/day)$^{-1}$. This yields the risk of 5.0×10^{-4} or 5 in 10,000.

Thus, for 10,000 people who were exposed by drinking water and eating fish from the river, 5 additional persons will come down with cancer in a period of 70 years. To place this in perspective, during the same period, about 1250 people out of 10,000 will die of cancer from other causes.

With the new control technologies in place, the discharge by 1985 had fallen to 3 pg/L for an estimated stream concentration of 3/30 or 0.1 pg/L. At this level, the risk from drinking water and eating fish from the river had fallen to 8 in 100,000. This is still outside the guidelines set by the state.

Conclusion

Before discussing the conclusions, three items of general interest will be mentioned. The first is an example of the heuristic of representativeness described in Chapter 1.

In March 1983 (EPA, 1983) Regional Administrator V. Adamkus of Chicago released a report that concluded ".... that Dow has extensively contaminated their facility with dioxin and has been the primary contributor to contamination of the Tittabawassee and Saginaw rivers and Lake Huron." Since Dow has been represented in the media as one of the largest producers of Agent Orange then it became very easy to associate the company as a major contributor of dioxin to the environment. The image fit; therefore, the conclusion that Dow was a major dioxin polluter must be correct. It is in this fashion that risk perceptions are made by the public and, once formed, are very difficult to change.

By performing a mass balance study on the Great Lakes watershed a loading rate of 600 to 800 g/year of dioxin would be required to maintain the observed water concentrations in Lake Huron (Neely, 1985). The 3 g/year from the Dow plant is less than 1% of the estimated total. If, as EPA charged, the Dow plant was the largest contributor, then the general conclusion must be that there are many small contributors to the total dioxin entering the Great Lakes. This would be in line with the next observation.

The second observation has to do with the proper release of new scientific ideas.

In the late 1970s a group of scientists of the Dow Chemical Company did some excellent work on the chemistry of dioxin formation. Unfortunately, Dow decided to call a press conference to release the information, rather than release it through the regular channels of peer review journals. The idea and conclusions that organic carbon plus chlorine, oxygen, and high temperature produce dioxin was sound. The media played up the idea and headlines emerged talking about the chemistry of fire and that dioxin had been around since Prometheus. There was much ridicule of Dow and it took a long time before the idea was finally accepted that dioxin could be formed in a variety of places such as incinerators. The length of time to have this idea accepted would have been shorter and the public perception would have been better if the conclusions had emerged naturally through the normal scientific channels (Bumb et al., 1980) rather than through a press release.

Finally, the actions that Dow took to resolve the socioscientific issue of dioxin contamination are a case study in how such disputes should not be handled.

One has only to examine the media articles during the early months of 1983 to realize that the company was being confrontational in attempting to solve their public relation problems. Dow was ill served by this strategy and would have achieved more satisfactory results by applying the principles discussed in Chapter 8. The company did not appreciate the real concern that people had with dioxin contamination. By its action, an impression was created that something was wrong. The public would have felt more confident that the perceived risk was small to nonexistent with a more open-door policy. Fortunately, the company learned from this experience and the handling of similar problems has much improved.

Besides the above observations, there are a few other conclusions that can be made. Specifically, the study shows the questions that need to be asked and how the answers can be used to generate a risk assessment. In general the flow of events follows the risk assessment procedure discussed in Chapter 1 (NAS, 1983).

1. The nature of the toxicological hazard needs to be identified. This includes analyzing the chemical and physical properties to see how the chemical will behave in the environment.
2. An experimental dose response curve for the agent must be established.
3. Using the response curve extrapolation must be made to either the predetermined risk level (i.e., 1/100,000) or a safe level (1/100 of the NOEL). A judgment decision must be made about what extrapolated dose level is acceptable.
4. The exposure model for estimating the water concentration must be examined. Questions need to be asked about the suitability of the model for the location under study. For example, how realistic is the assumption that the population in the valley drinks 2 L of water/day and eats 0.1 pound of fish/week (6.5×10^{-3} kg/day) for 70 years? What is a better model for the location in question?
5. The key numbers need to be examined for reliability and credibility. Again using the dioxin story one key number is the BCF. The value of 51,600 is very high and deemed by some experts to be based on faulty data. By using the value of 18,540, (a number that appears more suitable according to some investigators), the RSD would be increased.

Finally, each individual must take all the information and formulate a risk assessment that is compatible with his or her own set of values.

CASE STUDY 2: RADON

Historical Background

Marie Curie, a Polish-born French chemist, and her husband, Pierre, announced the discovery of a new element in 1898. They had separated a very radioactive mixture from pitchblende, an ore of uranium. The Curies took four years to obtain the pure material from this mixture, which they called radium. For their discovery, they received the Nobel Prize in chemistry. Unfortunately, because of their high exposure to radioactivity, they later succumbed to cancer.

What is this mysterious reaction known as radioactivity? Chemically it is the process during which the nucleus spontaneously disintegrates, giving off high-energy fragments. Natural radioactive elements emit three kinds of particles called alpha, beta, and gamma. Alpha consists of the helium-4 nucleus (helium with two protons and two electrons). Beta is similar to an electron while gamma is electromagnetic radiation.

The Curies were not the first casualties of cancer due to radioactivity. However, their case was probably the first where radiation was directly implicated as the cause of death. Earlier in the 16th century miners who worked in the mines in the Erz mountains, a region between Germany and Czechoslovakia, were dying from what was called "mountain disease" later identified as lung cancer. Subsequent knowledge showed the cancer arose from being exposed to radiation from the uranium ore. The epidemic in the mines was the first real recognition of this disease that is presently killing over 100,000 Americans a year (Bureau of the Census, 1982). However, cigarette smoking among Americans and not exposure to radiation is the main factor in 80 to 90% of these deaths (Fisher et al., 1988).

While studies relating radioactivity to lung cancer were being investigated throughout the world, the problem did not receive national recognition in the U.S. until Christmas in 1984. At that time an engineer by the name of Stanley Watras was employed in a partially constructed nuclear plant in Pennsylvania. Upon entering the plant in the morning he triggered the monitors designed to identify workers exposed to excess radiation. This event was most surprising since the plant was not yet operational. The discovery was made that his exposure came from a high concentration of radon[6] in his house where the level was a huge 2700 pCi/L.[7] This was far higher than most miners had ever experienced. The media attention given to this event helped make radon a household word.

From all of this attention the EPA not only investigated the radon levels in Pennsylvania, but also attempted to decide how significant the problem was nationally. This case study will investigate the potential risk faced by the nation's homeowners who are inadvertently being exposed to radiation.

Risk Assessment

Hazard Identification

The three main topics to be discussed are the chemical and physical properties, toxicological effects, and routes of exposure. Together these form the basis for describing the hazard from radon exposure.

---

[6] Radon is the first disintegration product from radium and is related to the noble gases such as neon, argon, etc.

[7] These dose levels will be described in a later section; suffice it to say that 2700 pCi/L is indeed a high dose.

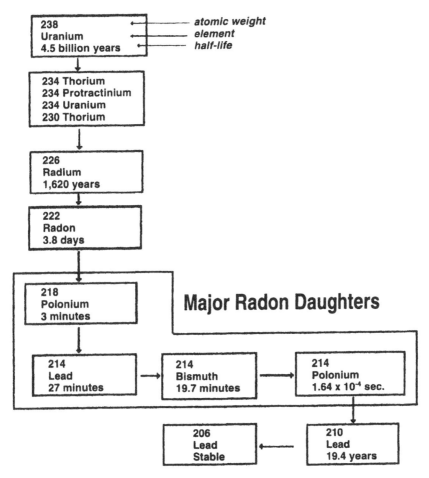

Figure 4.   Diagram illustrating the decay of uranium to the stable isotope
of lead.

Chemical and Physical Properties — Chemically, radon belongs to the group
of elements known as the noble gas series. These gases are helium, neon,
argon, krypton, and radon. The main characteristic of the group is their inertness
or unreactivity. This follows from the observation that the outer electron shells
are filled. Therefore, there is no tendency for these atoms to either accept or
donate electrons in a normal type chemical reaction. However, radon does
undergo decay, a process in which the nucleus spontaneously disintegrates,
giving off radiation. The process is further illustrated in Figure 4, where the
radioactive decay series for uranium is illustrated. Uranium-238[8] disintegrates
spontaneously over a period of billions of years to form radium-226 ($_{226}$Ra).

---

[8] 238 represents the atomic weight of the element and is usually represented as $_{238}$U.

This slow decay rate is obviously the reason there is still radioactive uranium ore present in the earth's crust. The rate now begins to speed up as radium emits alpha and gamma fragments with half-lives[9] of 1620 years to yield radon-222 ($_{222}$Rn).

Radon is an inert gas above $-61.8°C$ with a half-life of 3.8 days before decaying into the four "radon daughters" (Figure 4) that are solid and positively charged. These particles adsorb to the surface of dust particles, the lining of the lung, or any other charged surface such as the particles in tobacco smoke. Besides being adsorbed, the daughters have short half-lives and disintegrate releasing alpha, beta, and gamma fragments, ending with the formation of the stable isotope lead-206 ($_{206}$Pb).

The units used to express exposure are different from normal chemical reactions. They are different because the radiation is important, not the actual amount of chemical. Thus, the units show the strength of the emitted radiation.

Units of Radiation

The curie (Ci) is the SI unit for the rate of radioactive decay and is the amount of nuclide that undergoes 3.7×10^{10} disintegrations/second. The latter unit (disintegration/second) is a Bequerel (Bq). Therefore, 1 Bq = 1 disintegration/second and 1 Ci = 3.7×10^{10} Bq.

A pico Curie (pCi) is a more convenient term for expressing the rate of disintegration. From Chapter 2, the prefix pico (p) is 10^{-12}; consequently, a pCi will be equal to 3.75×10^{-2} Bq. It is customary to report the concentration in terms of pCi/L.[10]

Working level (WL) is equivalent to 200 pCi of radon/L (7400 Bq/m^3). This becomes a convenient one-parameter measure of the concentration of radon progeny in air. By combining the WL with time the working level month (WLM) was introduced so that both duration and level of exposure could be considered. WLM is the sum of the products of the WL times the duration of exposure during some specified period. The unit WLM is equal to 170 WL hours, which corresponds to an exposure of 1 WL for 170 hours (approximately one working month).

Using this definition, the risk experienced by Watras when he was exposed to 2700 pCi/L (100,000 Bq/m^3) will be examined. To assess the risk, the duration of exposure is needed. For example, 2700 pCi/L is roughly equivalent to 14 WLs. If this lasted for 5 hours/day for a year, then the exposure would

---

[9] Half-life is the length of time required for half the reaction to occur. Thus, in 1620 years half of the original radium is still present. Half-life is further described in Appendix II.

[10] Many countries report concentrations in terms of Bq/m^3. These units will be given in parentheses.

Table 4.   Summary of Number of Lung Tumors to Various Exposure Periods of Radon for Rats

Exposure (WLM)	Duration of exposure (weeks)	Exposure rate (WLM/week)	% Tumors
50	~10	2–8	2.9
290	~32	9	10
1470	4	370	25
2100	6.5	320	43
3000	8	370	43
4500	12	370	73

Data from Chameaud et al. (NRC, 1988).

be 150 hours/month or about 12 WLM for that year. The next sections will describe how the risk from different exposures can be determined.

Toxicology — The main toxicological property that is of interest is the effect of long-term inhalation of radon gas. Many different animal studies have been conducted over the last 50 years and have been summarized (NRC, 1988). Table 4 and Figure 5 show typical results from an exposure experiment. Several observations were made from these and other experiments (NRC, 1988).

- The latent period for the tumor development increased with decreasing cumulative WLM. In other words as the exposure decreased it required a longer period for the tumors to develop.
- The incidence of lung cancer for the same cumulative exposure increased as the duration of the exposure period was lengthened. In other words, a small repetitive dose over time is more effective in initiating tumors than a single large dose.

Routes of Exposure — Since radon is a gas inhalation is the most important route of exposure and becomes a very efficient delivery system for radiation to the lungs (Hansen, 1989). Radon reaches the air passages and deposits quickly on the surface of the lung. Since the daughters are very reactive they serve as the primary source for irradiating the cells—a first step in the formation of lung cancer. While the range of the highly ionizing alpha particles is very short, less than 70 μm (the thickness of a piece of paper), the epithelium in which the stem cells are located is only 40 μm thick. Consequently, the stem cells in which cancer is initiated are right in the target range for this radiation.

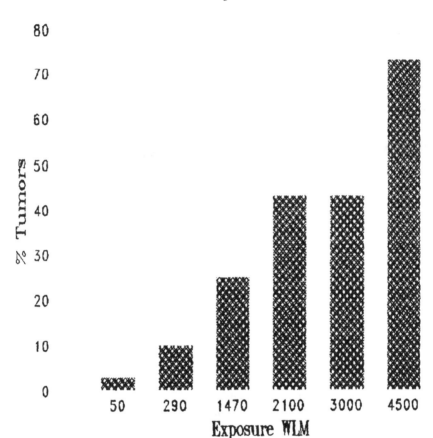

Figure 5.   Graph of the exposure and tumor response from the data in Table 4.

Dose Response

By the 1940s, radon was widely suspected to be the cause of the lung cancer in the miners working in the Erz mountain region. The rock being mined had a high concentration of uranium, and uranium was a known source of the radioactive gas radon. However, this was all circumstantial as the Erz mountain regions were notorious for a wide range of chemicals having the potential to cause cancer. In the 1950s a decision was made to study U.S. uranium miners, where the presence of radon could be easily monitored. This situation allowed for a controlled study in which a group of miners exposed to radon could be compared to a group not exposed. After 10 years of record keeping the conclusion was reached that the U.S. uranium miners had a higher incidence of lung cancer than the control group. Moreover, the symptoms of the exposed miners were comparable to the miners in the Erz mountains.

These results caused immediate action to be taken both in the U.S. and in Czechoslovakia to improve ventilation and other working conditions. In both places, as the radon levels decreased, there was a corresponding decline in the incidence of lung cancer. In search of more evidence, scientists investigated several different mines. In all cases where radon levels were high, the miners were found to have lung cancer, roughly in proportion to their exposures to radon. In addition, the observation was made that cigarette smoking enhanced the effects of radon exposure. A new study in 1976, from the mines in Czechoslovakia, suggested that, while the radon levels were much lower than previously noted, there was still an above-normal incidence of lung cancer. This higher than normal incidence occurred among miners who began working after ventilation improvements took place in the 1940s. The excess was half as large as earlier and was reduced in direct proportion to the reduction in radon levels. All the data from the miner groups are consistent, in giving roughly the same cancer risk per unit of exposure to radon. A final test compared radon's effects with those of other radiation sources. When the radiation dose from radon was expressed in the same terms as exposure from the atomic bomb or X-ray, the lung cancer risks were roughly the same. With all these data radon-induced lung cancer has become the most quantified health effect of radiation. By measuring the exposure to radon, the risk of developing cancer can be estimated.

There have been many retroactive studies of mine workers exposed to radiation from radon. Table 5 shows the results of one such analysis. From this data 22-WLM exposure exhibits no significant difference between the observed and the expected number of lung cancers. This latter number is the number of tumors that occur in a normal population of people not working in the mine. The same observation was made in most of the studies ranging from Czechoslovakia to Colorado to Canada. Generally, once the exposure

Table 5.   Observed and Expected Lung Cancer Deaths by Cumulative WLM among Ontario Uranium Miners (NRC, 1988)

Mean cumulative exposure (WLM)	No. of lung cancer deaths		No. of person years at risk
	Observed	Expected	
3[a]	14	11.7	51,356
12	13	17.2	61,823
29	15	11.0	38,751
53	13	7.0	23,313
98	12	6.0	17,435
200	15	4.1	10,208

[a]No explanation was given for the higher than expected lung cancer cases in the low-exposure group.

Table 6.   Characteristics of Domestic (Home) and Mine
Environments

Domestic	Mine workers
Diverse group ranging from very young to the very old	Homogeneous group composed of males between age 20–60
Low concentration of dust particles in the air	High concentration of dust particles in the air
More sedentary	Greater exertion—more physical

exceeded 40 to 50 WLMs, the number of excess deaths due to lung cancer began to increase.

The challenge, as pointed out by the committee on the biological effects of ionizing radiation (NRC, 1988), was to use all of the epidemiological and experimental data to address the issue of exposure in domestic environments. The two situations are different, as shown in Table 6.

Dust particles in the air receive the deposition of radon progeny. These particles taken in during breathing create a much stronger source of radiation.[11] In addition, the greater physical exertion by mine workers causes an inhalation of a greater volume of air. Both characteristics (dust concentration and heavy exertion) enhance the potential for injury to miners as opposed to home dwellers, from identical concentrations of radon. These types of differences make it very difficult to take the statistics from mine surveys and use them in the domestic situation. However, without actual data on homes, such translations are being made.

Exposure Assessment

There is no doubt that inhalation is the main route by which humans are exposed. The next two sections will discuss the sources and the exposure.

Source — The normal decay of uranium and thorium produces radon, a naturally occurring radioactive gas. Table 7 illustrates the average concentration of these two minerals found in soils (Boyle, 1988). Soils derived from granite and from shale have a high probability of containing elevated levels of radon. Areas containing the greatest activity are shown in Figure 6. Since these soils may contain radon, the gas will diffuse along a concentration gradient and because of a short half-life[12] less than a meter is required before most of the activity has been lost. A drop in the barometric pressure and an increase in

---

[11] This is one of the reasons for the synergistic action of cigarette smoking on radon-induced lung cancer.
[12] Half-lives are discussed in Appendix II.

Table 7. Average Soil and Rock Concentration of Uranium and Thorium

	U (mg/kg)	Th (mg/kg)
Rocks		
Igneous		
Silica (granites)	4.7	20.0
Intermediate (diorites)	1.6	8.0
Mafic (basalt)	0.9	2.7
Ultramafic (dunites)	0.03	6.0
Sedimentary		
Limestone	2.2	1.7
Carbonates	2.1	1.9
Sandstones	1.5	3.0
Shales	3.5	11.0
Mean value in earth's crust	3.0	11.4
Soils		
Typical range	1–4	2–12
World average	2.0	6.7
Average specific activity (pCi/kg)$^{-1}$	670	650

the wind speed or in temperature all cause the mass flow of radon to increase out of the soil. These processes are responsible for the flux over a larger distance. The entry rate into the house depends on the concentration of radon in the soil, the gas permeability of the soil, pressure gradients, and the structural integrity of the house. Loose sandy soil promotes radon transfer, whereas clayey, wet, or frozen soil inhibits gas flow. Once inside the house radon can accumulate, especially when the warm air inside rises and pulls up cooler air from the lower floors.

Exposure Assessment — A typical range for radium and uranium concentrations in soil produces 200 to 1000 pCi/kg (7.4 to 37 Bq/kg) which disintegrates to radon gas and percolates to the surface. At 1 m above the soil the level of radon falls to about 1 pCi/L (37 Bq/m^3) (Boyle, 1988). Obviously, this will vary considerably with the climatic conditions. The EPA estimates that the average outdoor level is about 0.2 pCi/L (7.4 Bq/m^3) (EPA, 1986). This type of situation will occur only in the regions of the country containing the ores of uranium and thorium (Figure 6).

By seeping through cracks in the foundation of homes, the radon becomes trapped and may build up to a much higher concentration than exists on the outside. Figure 7 illustrates some common entry points (EPA, 1986). By

**Figure 6.   Map showing areas with a potentially high radon concentra-
tion (Boyle, 1988). Reprinted with permission from *Environ.
Sci. Technol.*, 22, 1397, 1988, Copyright by the American
Chemical Society.**

conducting statewide surveys, the EPA has concluded that there are many
houses in the country with a high concentration of radon (greater than 20 pCi/
L, Hansen, 1989). In addition, one house can have a high level while the one
next door will be low. Based on this type of information, the recommendation
is to monitor basements located in active regions (Figure 6).[13]

Risk Characterization

As pointed out, the difficult challenge is to take the epidemiological work
on the miners (NRC, 1988) and apply it to the domestic environment. The
EPA has done this, and has made a series of recommendations that are based
on the data reproduced in Table 8 (EPA, 1986).

The rule of thumb given by the EPA is that, if room monitoring shows a
level of 4 pCi/L (148 Bq/m³), then the space needs to be ventilated and the
levels rechecked. Usually, ventilation (i.e., open the windows and blow out
the air with a fan) will clear the room and lower the concentration of radon
to a more acceptable level. From the data available 0.02 WL (4 pCi/L) may
be interpreted as follows:

---

[13] There are a number of reliable "do-it-yourself" kits on the market that make such testing
easy to accomplish as well as being economically feasible (EPA, 1986).

Common Radon Entry Points

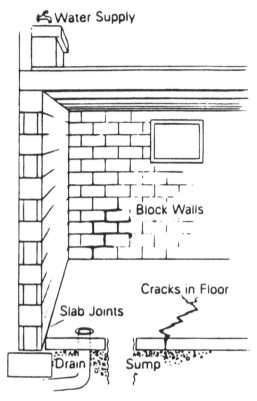

Figure 7. Common radon entry points in a home basement. (Reprinted from EPA publication EPA-86-004, 1986.)

1. 0.02 WL for a 24-hour day will yield 0.02 WLM/week of exposure.
2. If an individual spends a lifetime (70 years) in this environment he will achieve 52 × 70 × 0.02 for a total of 72 WLM of exposure. At this level of exposure the data suggest that the risk of contacting lung cancer is greater than the normal background rate (Table 5).

From the above interpretation, the first consideration is to decide the length of exposure and act accordingly. For example, not too many people spend a lifetime in one room. How much time is spent? One conclusion is that 4 pCi/L should be considered a cautionary number and should not cause panic.

Conclusion

The EPA has recommended that every house in the U.S. be monitored and that remedial action be taken when the radon level is at or above 4 pCi/L.

Table 8.   Radon Risk Evaluation Chart

pCi/L	WL	Estimated lung cancer deaths due to radon/1,000	Comparable risk
200	1	440–770	60 × nonsmoker risk 4 packs/day
100	0.5	270–630	20,000 chest X-rays/year
20	0.1	60–210	2 packs/day
10	0.05	30–120	5 × nonsmoker risk
4	0.02	13–50	200 chest X-rays/year Nonsmoker risk
1	0.005	3–13	Average indoor level
0.2	0.001	1–3	Average outdoor level

According to one estimate (Nero, 1992) this policy would result in the expenditure of $10 to $15 billion. The cost could go even higher if the public accepted the view expressed in certain EPA quarters that any level of radon is a concern. Another factor that could elevate the cost is that in 1988 Congress passed the Indoor Radon Abatement Act, establishing a long-term goal to reduce indoor concentrations to outdoor levels. This would entail the application of very complex and expensive technology with no assurance that it would be successful in preventing radon from entering houses. Fortunately, the act did not suggest a deadline.

The EPA policy, by focusing on the total country, has ignored what is a real problem in certain locations. Scientists have discovered that 6 to 7% of houses in the U.S. have levels exceeding 4 pCi/L while 0.1% (50,000 to 100,000) have levels above 20 pCi/L. This latter figure describes situations in which exposure is above the limits set for workers who handle radioactive materials (Nero, 1992).

A more sensible strategy (Nero, 1992) included the following elements.

1. A better education of the public is required to place radon exposure in the proper perspective. The present plan of focusing attention on every house is nonproductive and confuses the citizens about what is the real problem.

2. An effort needs to be made to find the small percent of homes where the radon levels are >20 pCi/L. This can be achieved through proper use of geological information and developing a radon map of the country. Once this is done a monitoring program can be begun and proper remedial action can be taken.

Such an integrated program would address the indoor radon problem in a meaningful manner. Houses that have a problem would be identified and corrective action taken. This would lower the exposure to radon and thus reduce the risk of lung cancer.

CASE STUDY 3: OZONE

Historical Background

To understand the political and scientific problems associated with the stratospheric ozone, the history and evolution of the refrigeration industry needs to be examined. During the early part of the 20th century there was an increasing demand for new technologies to protect food and provide air conditioning. Recall that refrigerants are liquids that are easily vaporized and in the process extract heat from their surroundings. The vapor circulates, enters a compressor, condenses to a liquid, and the process is repeated. The first refrigerant was discovered over 150 years ago by an English scientist who changed ammonia gas to a liquid by applying pressure and lowering the temperature. As the pressure was released, the ammonia liquid boiled off and caused a cooling to take place. This mechanism has formed the basis for modern refrigeration ever since. Over the years ammonia, sulfur dioxide, methyl chloride, propane, carbon dioxide, and ethane have all been used. The main material in use before 1930 was sulfur dioxide. While this liquid was efficient, it suffered from some serious drawbacks. For example, the gas was toxic, had a very pungent odor, and when mixed with water quickly formed sulfurous and sulfuric acid, both of which were extremely corrosive. The toxicity of these various gases and the need for large expensive compressors kept mechanical refrigeration from making headway with homeowners. That is why the synthesis by T. Midgely, Jr. in 1931 of dichlorodifluoromethane represented a major breakthrough for the industry. While working for the Frigidaire division of General Motors, he developed the molecule as the perfect alternative to the other gases on the market.

The DuPont Chemical Company picked up on this discovery and quickly began producing a whole series of chlorofluorocarbons (CFCs) under the name of Freons.[14] A few of the structures and some key properties are shown in Table 9. Freons, as a group, were easily vaporized and the latent heat of vaporization was great enough for them to cool down a closed chamber. Equally important, they were nontoxic, nonreactive, and appeared to be ideal agents for refrigeration.

Besides being used for refrigeration, the Freons found wide application in aerosol cans. The property that made them valuable was that under pressure and normal temperature the chemicals were liquid. Once the pressure was released the liquid changed to a gas, carrying with it the contents of the can.

---

[14] Freon is a registered trademark of the Dupont Chemical Co.

Table 9.   Properties of Some Key Freons

Chemical	Structure	Bp (°C)	Pressure[a] (kPa)
Freon 22	$CHClF_2$	−41	1,101
Freon 114	$C_2Cl_2F_4$	3.5	151.5
Freon 12	CCl_2F_2	−29.5	642
Freon 11	CCl_3F	23.5	24.6

Note: 101.3 kPa = 1 atm.

[a] This is the pressure required to turn the gas into a liquid under the operating conditions of the refrigerator.

Many convenience packages were manufactured containing Freons mixed with items such as paint, deodorants, hair sprays, etc. After being released from the can, the CFCs entered the atmosphere where they were quickly dissipated and forgotten. The additional properties of being nonflammable, nontoxic, and nonreactive were equally beneficial in this application. This use of the CFCs quickly surpassed the use as a refrigerant as shown in Figure 8.

The production of the CFCs began growing in an exponential manner shortly after their introduction in the 1950s, as Figure 8 indicates. The growth may be easily appreciated from the production figures for one member of the group. CCl_3F (Freon 11) started from 2.97×10^{10} grams (33,000 tons) in 1957 and grew about 18% a year reaching 3.67×10^{11} grams (405,000 tons) in 1973. Scientists from the University of California (Molina and Rowland, 1974) began wondering where all these materials were going. They knew they were entering the atmosphere, but what happened to them after that? Their early investigation was based on a simple mass balance calculation. By knowing how much CFCs had been produced they could compare that number with the amount that was present in the atmosphere from analytical measurements. An example of this type of analysis is reproduced in Appendix II.[15] The results show that most of the CFCs that had been made up to and including 1974 were still floating around in the troposphere. What would be their final disposition? Molina and Rowland concluded that the materials would eventually enter the stratosphere (Rowland, 1989) and (Molina and Rowland, 1974)[16] and be attacked by high-energy ultraviolet radiation. What would be the impact of the ultraviolet radiation on the CFCs? What would happen to the critical ozone layer? Finally, what would be the consequences of destroying this layer? While trying to answer these questions a better understanding was gained of the chemistry and physics of this important part of the environment. With this knowledge a greater awareness of the world was developed. This in turn has

---

[15] An understanding of this calculation requires the use of simple algebraic relations. However, being able to work the problem is not necessary for an understanding of this case study.

[16] For a description of this important area of the atmosphere, see Chapter 5.

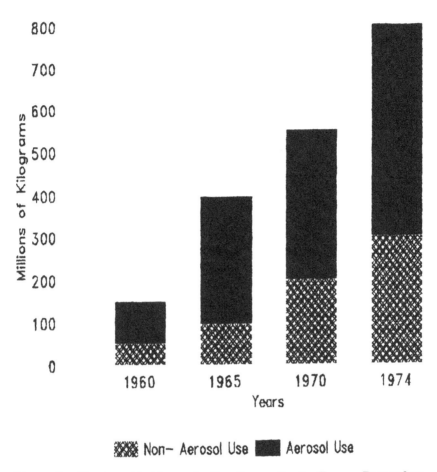

Figure 8. Total worldwide production figures for the Freons. Data taken from the Chemical Manufacturing Association.

led to a better definition of the potential risk of continually adding volatile halocarbons into the atmosphere. The next section will deal with an assessment of this risk.

Risk Assessment

The outline used in the previous case studies and described at the beginning of this chapter will serve as the basis for developing the following risk assessment.

Hazard Identification

This study is different from the previous examples since the chemicals under investigation, i.e., the CFCs, are nontoxic. In fact, this was one reason they became so widely used. Consequently, to examine the hazard from the continued use of the Freons, the area that needs investigation is their impact on the stratosphere. Thus, the section on chemical and physical properties will be devoted to stratospheric chemistry and the toxicological section will be concerned with the influence of increased ultraviolet radiation on biological systems.

Physical and Chemical Properties — The chemistry will deal with the series of reactions that create and destroy ozone in the stratosphere. Recall from Chapter 5 that ozone is formed by the reaction of O_2 and O. The destruction mechanisms that have been studied are

$$O + O_3 \rightarrow 2O_2 \qquad 20\%$$
$$OH + O_3 \rightarrow HO_2 + O_2 \qquad 10\%$$
$$NO + O_3 \rightarrow NO_2 + O_2 \qquad 69\% \qquad (5)$$
$$Cl + O_3 \rightarrow ClO + O_2 \qquad 0.5\%$$
$$\text{loss to the troposphere} \qquad 0.5\%$$

The percentages estimate how much each reaction contributes to the loss of ozone. These reactions are all natural mechanisms which bring the ozone layer into a steady-state concentration.

From the perspective of this chapter one important reaction is the interaction of chlorine with ozone. Of equal significance is the effect that man's activities are having on increasing the chloride burden in the stratosphere. The major source of the natural chloride is methyl chloride, a volatile chemical which originates in seawater and is eventually transported to the stratosphere by diffusion. The recent introduction of Freons and other volatile halocarbons is now adding to the chloride burden and causing further destruction of ozone. In its summary, the National Academy of Sciences (NAS, 1979) estimated that CFCs contributed the same amount of destruction as methyl chloride, i.e., less than 1%. However, if the growth of CFCs were to continue at the 18% level then obviously more destruction would occur in future years. In fact the reduction in ozone was predicted to be 7.5% at steady state (reached in the year 2075) for a constant 1973 release rate. Considering the uncertainty in the calculations, the percent reduction was bracketed between 2.4 and 15%. By 1982 NAS updated its estimate to 5 and 9% (NAS, 1982). Latest results also suggest that the CFC released to date should have reduced the total ozone column by less than 1%. The difficulty in analyzing the ozone layer and

Ozone concentrations fluctuate widely over time

% deviation, global average total ozone (ground-based data)

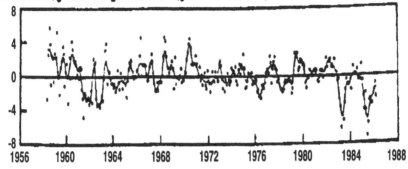

Figure 9.   Graph showing the percent deviation on a global average for the total ozone in the stratosphere. The dots are the actual data and the solid line is the trend.

establishing a trend line is shown in the data in Figure 9 (Ember et al., 1986).[17] In other words, the validity of the model cannot be established through monitoring since the natural variability in the stratospheric ozone exceeds the 1% level suggested by the model. However, it appears certain that the continued release of CFCs will cause a buildup in the concentration in the lower atmosphere and eventually reduce the concentration of stratospheric ozone. The consequences are at least twofold.

1. The continued buildup of the CFCs in the lower atmosphere will contribute to the greenhouse effect (Chapter 5). At the present time (Ember et al., 1986) it is estimated that the CFCs contribute an amount equivalent to about one fourth of the effect of carbon dioxide itself. This question will be addressed later in the chapter.
2. The destruction of the ozone shield will allow a greater intensity of ultraviolet radiation to strike the surface of the earth. The major concern is that the increased intensity might cause a corresponding increase in skin cancer. The potential risk of this happening will be discussed below.

Toxicology — Before looking at the toxicological data, recall that overexposure to sunshine causes two types of cancer, nonmelanoma and melanoma. The

---

[17] This graph is reprinted with permission from *Chem. Eng. News.* Nov. 24, p. 31, 1986, and G. C. Reinsel, University of Wisconsin. Copyright by the Americal Chemical Society.

first type or nonmelanoma is the most frequently detected cancer in humans and the most successfully treated of all cancers and rarely leads to death. Melanomas, on the other hand, are life threatening and need to be prevented if possible.

Most of the available data on the effects of ultraviolet radiation on humans comes from epidemiological studies rather than through direct experimentation. The section dealing with dose response will illustrate this observation in greater detail. In what follows, only a few of the laboratory experiments will be described. For example, investigations have been conducted on the response of bacterial systems to UV radiation. While these systems differ from human cells, the fundamental biology and genetics are analogous. From such experiments the observation is made that the bacterial DNA absorbs the energy from the short wavelengths. In the process the absorption causes mutation and death of the bacteria. Other studies using mice have shown that monochromatic ultraviolet radiation can cause changes in the animal's immune response. These and other types of studies show that alterations similar to the above may occur in human cells leading to the formation of cancer (NAS, 1982).

Route of Exposure — Investigations have shown that there must be direct contact of the skin with the ultraviolet radiation before an effect is observed. Consequently, absorption through the exposed skin becomes the main route of exposure (NAS, 1976).

Dose Response

Information provided in 1982 (NAS, 1982) strengthened the hypothesis of a causal relationship between nonmelanoma skin cancer and ultraviolet radiation. The data also confirmed the protective effect that skin pigmentation afforded the black population. Predictions of the increased incidence in nonmelanoma skin cancer due to reduction in stratospheric ozone are shown in Table 10 (NAS, 1982). The model used to make these predictions was based on actual observations from different latitudinal locations. In this manner a relation was established between incidence and intensity of ultraviolet radiation. The results of this model were then applied to the actual data from Minneapolis and Dallas using a hypothetical reduction in stratospheric ozone with the results as shown in Table 10.

The evidence for a similar relation between the incidence of the more deadly melanoma skin cancer and radiation is not quite so clear-cut. However, there does appear to be a latitude dependence as the data in Table 11 suggest (NAS, 1976). The closer to the equator, the higher the mortality and the higher the radiation. NAS concluded that the relation of solar radiation to melanoma

Table 10. Percent Increase in Skin Cancer for U.S. White Population with Reduction in Stratospheric Ozone for Two Locations

	% Increase in skin cancer for ozone reduction of	
	5%	10%
Minneapolis		
Male	5.6	11.2
Female	4.4	8.9
Dallas		
Male	8.4	16.9
Female	6.8	13.6

Note: These figures are for basal cell skin cancer. Squamous cell cancer was slightly higher.

should be taken as a health hazard of significant size, and that society should act accordingly. Society's response is seen in the increase in sales of sunscreen lotions and the observation that people are more cautious about exposing themselves to direct sunlight.

Exposure Assessment

Biological organisms are constantly being exposed to ultraviolet radiation. The question that needs answering is what would be the increase in exposure if a decrease occurs in the stratospheric ozone layer? If there were a resultant increase in exposure to UV radiation, would the effect be significant? The answer is found by realizing that the amount of UV reaching the earth's surface depends on the altitude. For example, in Denver the radiation level in July is about 12% greater than that in Washington, DC. While Denver and Washington are at the same latitude, the former is considerably higher in altitude. The "changing atmosphere" (Ember et al., 1986) contained a graph

Table 11. Mortality Data from Melanoma as a Function of Latitude

Latitude	Mortality (deaths/100,000 white population)
33–34	2–3
38–40	1–2
45	1

Data taken from "Environmental Effects of Chlorofluoromethane Release" (National Academy of Science, p. 87, 1976).

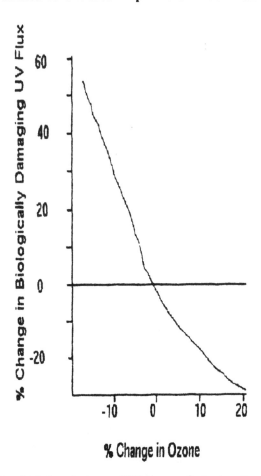

Figure 10.   Relation between UV flux and stratospheric ozone.

reproduced in Figure 10[18] showing the amount of increased ultraviolet radiation at the earth's surface that will result from a drop in the stratospheric ozone levels. NAS said that, if the production and use of the CFCs continued to grow at the 1973 rate, then by 2075 the decrease in ozone was estimated to be between 5 and 9% (NAS, 1982). Thus, Denver in July is comparable to what Washington would be like if there were an overall 5% change in ozone (Figure 11).

The question becomes, "Are there currently any detectable differences in the cases of melanoma between Washington and Denver?" From the available statistics it is impossible to detect such a change. In other words, if there were a 5% decline in total stratospheric ozone, a change in the incidence of mela-

---

[18] This figure is reprinted with permission from *Chem. Eng. News.* November 24, p. 33, 1986 and J. E. Frederick, University of Chicago. Copyright by the American Chemical Society.

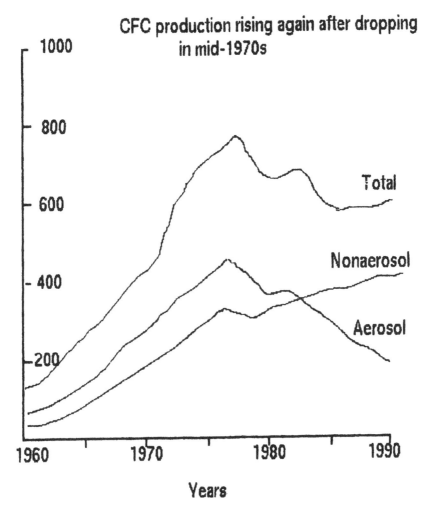

Figure 11.   **Worldwide production figures for CFC excluding Russia and China. Reprinted with permission from** *Chem. Eng. News,* **Nov. 24, p. 53, 1986. Copyright by the American Chemical Society.**

noma would not be seen in Washington. However, the incidence of the less deadly nonmelanoma would probably increase (Table 11).

Risk Characterization

The current data show that the present risk from increased exposure to UV radiation is small. The problem is attempting to predict what will happen 50 to 100 years from now. To make such predictions, assumptions have to be made about the starting conditions. To create a worst-case scenario, the amount of CFC introduced into the models is based on a continued 18% annual growth pattern for the next 50 years. Obviously, if that happened there would be a large concentration of chloride in both the troposphere and the stratosphere. Figure 11 shows what is actually occurring to the total worldwide production of CFC (Ember et al., 1986). The production peaked in 1975 and then dropped due to the decreased demand for propellants in the aerosol packaging industry. By 1983 the production began to rise due to the increased use of CFCs by countries other than the U.S. The data suggest that the early growth rate of 18% is not being sustained and the global impact will be less than suggested by the National Academy.

This was the situation in the late 1970s. With the restriction made by EPA in 1978 on the use of CFCs as aerosol propellants, the public interest in the issue calmed down. Most Americans were persuaded that the problem had been averted.

By 1986 the issue again was making the headlines and the evening news. This time the concern was related to a hole in the ozone layer that was being observed over Antarctica. This hole was first demonstrated by scientists from the British Antarctica survey team (Zurer, 1987). They discovered that in September and October the levels of ozone over Halley Bay, Antarctica, decreased annually. After an intensive study, scientists from many countries met in Colorado in 1988 to report their findings (Zurer, 1988). The conclusion was irrefutable; chlorine from the CFCs was largely responsible for the observed decrease in the ozone layer that took place every spring. Besides the chlorine, the unique dynamics of the Antarctica region play an important role. In the stratosphere, a stream of air known as the polar vortex circles the region in winter. Air trapped within the vortex becomes very cold and forms clouds. These polar stratospheric clouds in turn provide surfaces for heterogeneous reactions that convert inactive chlorine molecules to forms of chlorine that can catalyze the destruction of ozone. Once the vortex breaks up in mid spring the Antarctica air mixes freely with the stratospheric air at lower altitudes. This causes a dilution with a resultant further decrease in the ozone layer at mid latitudes. These investigations led to the signing of the Montreal accord in September 1987. The protocol signed by 24 countries of the United Nations and the European Community called for a freeze in the production of CFCs at the 1986 levels by July 1, 1989. Later cuts would reduce CFC consumption first by 20% in mid-1993 and 30% by 1998.

Besides being connected with the ozone layer these same gases have been identified with the "greenhouse effect" discussed in Chapter 5. Briefly, the global climate reflects the balance between incoming radiation from the sun and the ability of the atmosphere to reflect the infrared heat created at the earth's surface. A current concern is that a change in the earth's climate is being induced by human activities through the introduction of the "greenhouse" gases, carbon dioxide, methane, ozone, nitrous oxide, and a variety of synthetic chemicals such as CFCs, methylchloroform, and carbon tetrachloride.

There is now sufficient concern about the buildup of these last-named synthetic chemicals in the troposphere that the major producers have agreed to phase out the production by the early part of the next century (Zurer, 1989).[19] The problem of reducing carbon dioxide is much more difficult. In the summer of 1992 an international conference sponsored by the United Nations was held in Rio de Janeiro (Elmer-Dewitt, 1992). As might be imagined, agreeing to any type of fossil fuel reduction proposal was difficult. The third world countries could not understand why their plans for development should suffer in order to rectify a problem they did not create. The rich countries (in particular the U.S.) did not want to sign on to anything that would threaten their lifestyle or increase the cost of doing business. The European countries proposed cutting CO_2 emissions to 1990 levels by the year 2000. The U.S. refused to consider rigid deadlines. Finally, the proposal was sprinkled with enough ambiguities and voluntary goals that all countries agreed to sign. While the proposal is not perfect, it does represent a very important beginning to the resolution of what is truly a global problem.

Conclusion

It is fascinating to recall the incidents leading up to the present situation in the control of CFC releases. When the initial article appeared in *Nature* (Molina and Rowland, 1974), the subject became a media event. The notion that individuals could threaten the ozone layer miles above their heads by touching off spray cans caught the attention of the general populace. As more evidence began to accumulate, the sales of these spray cans began to drop voluntarily. Continued pressure on Congress caused the regulatory agencies to restrict the nonessential uses of the CFCs in the fall of 1978. Four years after the 1974 article, the regulations were in place—not to ban all the uses of CFC but to ban the nonessential uses. This was a prudent course of action, and the restrictions accomplished their purpose.

At this point the public interest all but disappeared. The dramatic announcement in 1986 that a hole in the ozone layer over Antarctica had been identified quickly refueled people's attention. Simultaneously came the realization that the problem was global in nature. Restriction of chemical use in the U.S. would not be enough. After many joint meetings between scientists and politicians from the United Nations, the previously discussed Montreal accord was signed (Zurer, 1988).

---

[19] As indicated in Chapter 7, the Dow Chemical Co. announced that it will speed up the process and will terminate the production of methylchloroform by 1995, *Chem. Eng. News.* April 20, 1992.

Currently, all parties including the producers recognize the problem and are working hard to resolve the issue. The producers are busy trying to find suitable substitutes for CFCs and have agreed to phase out production of the current CFCs. In addition, agreement has also been reached to stop the production of all other chlorinated solvents by the turn of the century (Zurer, 1993).

CASE STUDY 4: ALAR

Historical Background

Alar is the registered name of a product formulated by Uniroyal Chemical Company, containing the chemical daminozide. The chemical, discovered

$$(CH_3)_2-N-N-\overset{O}{\overset{\|}{C}}-(CH_2)_2-\overset{O}{\overset{\|}{C}}-OH$$
$$\underset{H}{}$$

Daminozide

in 1950, is a plant growth regulator and was first registered in 1968 for use in controlling vegetative and reproductive growth of orchard crops such as apples, cherries, nectarines, peaches, and pears. Daminozide affects flower bud initiation, fruit set, and maturity. Through these mechanisms the fruit (particularly apples) remain on the trees longer, allowing a greater percent of the crop to mature. In addition, the treatment prolongs the storage life of the harvested crop.

From 1968 to 1984 the product was used successfully for the above purposes without any perceived problem. In July 1984 the situation began to change. The EPA placed daminozide under administrative review. This review was triggered by a cancer study that had been performed in 1977 (Toth et al., 1977) at the Eppley Institute in Omaha, NE, on both daminozide and the hydrolysis product, unsymmetrical dimethyl hydrazine (UDMH).

$$(CH_3)_2N-NH_2$$

UDMH

In August 1985 the EPA announced the possibility of canceling uses of daminozide. As required under federal law, the EPA had to submit its toxicology and exposure data to a Scientific Advisory Panel for a peer review. The panel was highly critical of the Toth study (which was the main part of the EPA's toxicological package) pointing out that the dose of daminozide far exceeded the maximum tolerated dose (MTD). Chapter 4 described how such a dose can cause cellular stress. The animals' response to this increased pressure is to produce new cells. As a consequence the animals are at greater risk to events that can lead to cancer. Many toxicologists recommend that cancer testing protocols should restrict the dose to a level well below the MTD. When daminozide is fed at these lower doses, tumor development is not seen. Besides the use of high doses, the Toth study had several other flaws in the experimental design (Exton, 1989). These are interesting to examine in light of the discussion in Chapter 2. For example, (1) only one dose level was used as opposed to obtaining a dose response curve from several doses, (2) control animals were not run concurrently; in fact one group of controls was run at a different institution; (3) the animals fed daminozide were given extensive doses of antibiotics, while the control animals were given antibiotics only occasionally.

Based on the review of the Scientific Advisory Panel, the EPA on January 22, 1986, rescinded the proposed cancellation of daminozide. Simultaneously, they requested that Uniroyal conduct additional cancer and exposure studies. The company did these investigations and at the request of the EPA turned in a preliminary report before the experiments were finished. This will be discussed in greater detail in the next section. Based on these early results, the EPA (Moore, 1989) backed off their decision of January 1986 and indicated an intent to move quickly toward cancellation of the use of daminozide on fruit crops. Concurrently, the Natural Resources Defense Council (NRDC) was preparing a report using the same Toth studies that had been criticized by the EPA's Scientific Advisory Panel. This report was to form the basis for bringing public pressure to bear on the regulatory agencies and hasten the cancellation process. The report (NRDC, 1989), which had never received peer review, was used for a CBS "60 Minutes" segment. This aired on February 26, 1989. The program was designed to catch the attention of the American public and this they did with a great deal of success. "The most potent cancer-causing agent in our food supply" said reporter Ed Bradley standing in front of a skull and crossbones emblazoned on an image of big red apple. Bradley went on to cite the findings of ". . . a number of scientific experts" who argued that eating those apples over a lifetime could cause thousands of cancer cases. The identification of the "experts" was not made. In addition, there was no mention that many other toxicologists strongly disagreed with the NRDC conclusions.

In the following weeks, the morning and evening TV news programs contributed to the visibility of the issue by giving credence to many people, including actress Meryl Streep, who were requesting EPA to ban such pesticides from the food supply. Other media coverage included cover stories in *Time* and *Newsweek* (two stories in each magazine), and the Phil Donahue Show. In addition, there were multiple stories in the major newspapers across the country, three cover stories in *USA Today, People* magazine, and four women's magazines with a combined circulation of 17 million (*Redbook, Family Circle, Woman's Day,* and *New Woman*).[20] By this time Alar had become a national issue recognized by millions of people. In the face of this

---

[20] *The Wall Street Journal,* Oct. 3, 1989.

continuous bad publicity, Uniroyal on June 2, 1989, voluntarily halted all sales of daminozide and recalled all products that had been sold or were sitting on shelves waiting to be sold.

Obviously, with the termination of sales and production, the risk from exposure to Alar has ceased to be an issue. However, the question still remains, was there a risk? In the next section, the problem will be addressed using the tools that have been developed in this book.

Risk Assessment

The procedure of assessing the potential risk will be similar to the other case studies and will follow the usual outline.

Hazard Identification
Dose Response
Exposure Assessment
Risk Characterization

Hazard Identification

The topics that will be discussed in identifying the hazard are physical, chemical, and toxicological properties and the route of exposure.

Physical and Chemical Properties — Daminozide is a white, crystalline solid with very little odor and a melting point between 154 and 156°C. It is soluble in water, methanol, and acetonitrile, but insoluble in xylene and other aliphatic hydrocarbons.

Recent studies (see next section) showed that the hydrolytic breakdown product UDMH is the main carcinogen rather than daminozide. With this realization came an interest in determining to what extent UDMH can be formed during the normal processing of daminozide-treated apples. For example, in 1986 the U.S. Food and Drug Administration (FDA) conducted a survey to determine residues of daminozide and UDMH in fruits and fruit products (Saxton et al., 1989). Stored fresh apples were found to contain 0.6 mg/kg (ppm) daminozide and no UDMH. On the other hand, applesauce and apple juice, produced from apples under high-temperature conditions, contained not only daminozide at 0.8 to 1.1 mg/kg but also UDMH at levels of 0.04 to 0.06 mg/kg. A follow-up investigation (Santerre et al., 1991) showed that less than 10% of the daminozide was converted to UDMH under conditions[21] simulating the processing of apples to juice.

---

[21] These conditions included a step in which the solution was maintained at 100°C.

Toxicological Data — The acute LD_{50} in rats was determined to be 8400 mg/kg (NIOSH, 1980). Recall from Chapter 3 that a dose of this size is considered relatively harmless. Studies showed that daminozide was not metabolized when fed orally to either rats or dairy cows (St. Johns et al., 1969). In addition, 93 to 95% of the dose was excreted unchanged.

Cancer Studies — Mention must be made of the early cancer studies (Toth et al., 1977) since these are at the heart of the report by NRDC (NRDC, 1989). The studies were performed using a single dose of 150 mg per mouse per day. Assuming a typical mouse weighs 30 g, this corresponds to 5000 mg/kg/day.[22] With this high dose (recall that the acute LD_{50} in rats was 8400 mg/kg) about 72% of the animals developed some form of blood or lung cancer. In addition, many treated animals suffered from toxicity symptoms other than cancer. These studies were discarded by the EPA's Scientific Advisory Panel for the reasons cited earlier.

Subsequent studies have shown UDMH to be the active carcinogen rather than the parent chemical, daminozide (Moore, 1989). The question then becomes, how much UDMH is present in products containing daminozide? For example, in the studies of Toth based on feeding daminozide, the assumption is that the cancer was initiated by the contaminant UDMH. Furthermore, the dose of UDMH from feeding daminozide at 5000 mg/kg/day was estimated to be 29 mg/kg/day (Marshall, 1991). That 29 mg/kg/day of UDMH is above the MTD is partially substantiated by studies using turkeys (Simpson and Barrow, 1972). These workers fed UDMH mixed in the feed to turkeys for several weeks. There was no weight gain in the treated animals and at the end of the experiment there was histological evidence of liver damage. The amount used was reported as 45 mL of UDMH mixed with 50 lb of feed. The turkeys on the diet consumed 1 lb/week. By making the assumption that the density of UDMH is 1 g/mL and that the average weight of the turkeys was 5 kg, a dose can be estimated as 26 mg/kg/day. While turkeys are not mice, this dose, similar to the amount used in the cancer studies (Toth et al., 1977) only gives liver damage with no cancer being reported.

A summary of the more recent feeding studies using daminozide and UDMH appeared in the *Federal Register* (Moore, 1989) and is given in Table 12. These were performed by Uniroyal at the request of the EPA. Unfortunately, the results are from a study that was only half finished. The EPA was under pressure and felt that it could not wait another year for the final report.[23] Based on these partial findings, the EPA made a decision to ban all products containing daminozide. Consequently, the results of the final investigation sponsored by Uniroyal have never been published.

---

[22] 5000 mg/kg/day for a 150-pound person is equivalent to 350 g/day or 3/4 of a pound of daminozide per day.

[23] Note that the *Federal Register* notice appeared on February 10, 1989, and the NRDC/CBS documentary aired on February 26, 1989.

Table 12.   Incidence of Blood Vessel Tumors in Mice from a Feeding Study Using Daminozide and UDMH

| | Daminozide concentration in water (mg/L) | | | | |
	0	300	3,000	6,000	10,000
Male	0.12	0.02	0.06	0.06	0.21
Female	0.09	0.02	0.04	0.02	0.14

| | UDMH concentration in water (mg/L) | | |
	0	40	80[a]
Male	0/45(0)[b]	1/45(0.02)	11/53(0.22)
Female	0/43(0)	1/47(0.02)	8/51(0.16)

[a]The 40 and 80 mg/L in the drinking water converts to a dose of 11.5 mg/kg/day and 23 mg/kg/day in the mouse (Moore, 1989). It is assumed that a 35-g mouse drinks about 10 mL of water a day.

[b]The numbers in parentheses show the fraction of animals responding, i.e., 0 animals out of 45 had tumors for a response of 0.

As EPA showed (Moore, 1989), the tumors observed in the high-dose daminozide study were similar to the type present in the UDMH investigation. Low-level contamination of the daminozide could have contributed to the observed tumors. The high concentration (10,000 mg/L) in the drinking water converts to a dose of 2.8 g/kg. This dose is the same size as used in the Toth study (5 g/kg). Accordingly, the results are subject to the same types of criticism that the Scientific Advisory Panel leveled at the Toth study. Presently, there is no strong evidence that daminozide by itself is a carcinogen.

A 12-month study using UDMH in the drinking water at lower levels of 0, 1, 5, and 10 mg/L in male mice showed no increase in tumors because of the treatment. A comparable study using female mice at twice the concentration showed the same lack of effect. In a similar vein, a 12-month rat study using UDMH at a maximum concentration of 100 mg/L in water also showed no tumors (Moore, 1989). The only increase in tumors came from the high exposure of UDMH in mice as summarized in Table 12. As indicated earlier, a dose of 23 mg/kg/day (80 mg/L in water) is close to the MTD level that caused cellular toxicity in turkeys.

Route of Exposure — Obviously, the main route of exposure for humans is the eating of fruit and fruit products that have been exposed to daminozide materials. This is the reason that several "marketbasket surveys" have been conducted. These are studies in which randomly selected products are taken from grocery shelves and analyzed for daminozide and UDMH (Saxton et al., 1989, is one example of such a study).

Dose Response

The EPA attempted to use the data in Table 12 to establish a dose response curve (Moore, 1989). From examining the number of tumors at the low dose of 11 mg/kg/day, only 1 out of 45 animals elicited a response. To detect a reaction at the 2% level a sample of 1237 animals would be required (Chapter 2). Obviously, with only 45 animals it is impossible to make a statement[24] regarding the single tumor that was observed. The conclusion that must be drawn is that the present data are not suitable for constructing a meaningful dose response curve.

Exposure Assessment

Under the Federal Insecticide, Fungicide, and Rodenticide Act, pesticide residues are expected to occur in produce that have been sprayed. The level that occurs when the chemical is applied at a rate needed for pest control is called the "tolerance level." These concentrations are examined for their toxicity. If, in the judgment of the EPA, the amounts found cause health problems, the request for registration may be denied. The tolerance for daminozide was set at 20 mg/kg (Moore, 1987). This was based on actual crop field trials as well as marketbasket surveys. The trials were conducted in the apple-growing states of Washington, Michigan, New York, Pennsylvania, and North Carolina. Samples were collected on a variety of apples that included Red and Golden Delicious, McIntosh, and Jonathan. Treatment rates were 2.5 to 3.4 pounds of active ingredient per acre (2.7 to 3.8 kg/ha). Apples were harvested from 35 to 60 days after treatment. The residues of daminozide ranged from <1 to 10 mg/kg with an average for all samples being 4 mg/kg (Moore, 1987).

The actual levels found in the products by the time they reach the consumer will be considerably less. This was discovered to be true by the studies performed by the FDA and Uniroyal. They reported on separate marketbasket studies that showed an average concentration of daminozide on apples of 1 mg/kg, much less than the tolerance level of 20 mg/kg. Residues of UDMH in apples ranged from <0.001 to 0.027 mg/kg for an average of 0.007 mg/kg. As expected, the level in apple juice and applesauce was higher — 0.033 (0.112) to 0.044 (0.228) mg/kg, respectively. (The numbers in parentheses were the maximum values found.) The worst-case scenario for UDMH based on a tolerance of 20 mg/kg of daminozide would be about 1 mg/kg. This is derived from a conversion of daminozide to UDMH of 6.2% found in a processing study (Moore, 1987).

---

[24] The EPA in support of its decision to cancel the registration pointed out that the tumor was of a type not normally found in control animals. Consequently, it must have been initiated by the treatment.

Risk Characterization

In all animal studies that have been reported, there is no evidence that daminozide is a carcinogen. However, the hydrolytic break-down product UDMH does initiate tumors. The initiation occurs only with a dose that approaches the MTD. UDMH triggers tumor production when the cells are stressed and are beginning to multiply because of the cellular toxicity. Accordingly, UDMH may be classified as a promoter (Chapter 4). Further evidence for this conclusion is the observation that there is no evidence of tumor development when the dose is maintained below the MTD.

From the results reported by the EPA (Moore, 1989) no visible tumors in mice were found after dosing with 10 mg/L of UDMH in the drinking water. This will be considered the no observable effect level (NOEL). With the assumption that a 30-g mouse drinks 10 mL of water a day, the 10 mg/L is converted to a daily dose of 3.8 mg/kg/day. From the discussion on promoters in Chapter 4, this daily dose may be converted to a safe dose for humans by dividing by 100. Consequently, if exposure to UDMH is at or below 0.038 mg/kg/day, humans should be considered safe from the initiation of tumors by this agent.

Human exposure to the chemical is based on apple and apple juice consumption as documented by the U.S. Department of Agriculture (USDA, 1985). According to this study, children are at the greatest risk because of their liking for apple juice. For a typical child (<5 years old) the intake of juice is between 54 and 68 g/day. By contrast an adult drinks about 14 g/day (USDA, 1985). Assuming the highest contamination value found for UDMH in juice (0.044 mg/kg, Moore, 1987) an exposure of 0.0002 mg/kg/day is estimated from the following equation:

$$Exposure = \frac{0.068 \ kg \ juice}{day \cdot child} \times \frac{0.044 \ mg \ UDMH}{kg \ juice} \times \frac{1}{15 \ kg/child}$$

$$= 0.0002 \ mg/kg/day$$

(Note how the units cancel, leaving the dose in mg/kg/day)

Using a maximum concentration of 0.228 mg/kg, found for UDMH in apple products, increases the exposure to 0.001 mg/kg.

This is well below the exposure level of 0.038 mg/kg/day estimated to be safe. If the products containing daminozide are used according to the label directions, there should be no hazard. However, when EPA analyzed the same data (Moore, 1989) they concluded that the registration for Alar should be revoked. In addition, the Natural Resources Defense Council (NRDC, 1989) came to a more critical conclusion when they labeled the chemical UDMH as a very potent carcinogen. They estimated that over 6000 preschool-age children would die of cancer initiated by exposure to contaminated apples.

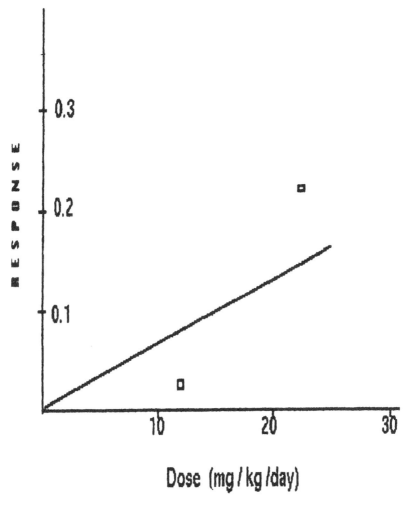

Figure 12. **Tumor response curve for the data on male mice fed UDMH at the indicated doses (Moore, 1989).**

To help clarify these diametrically opposed views, the next section will describe how the EPA and NRDC arrived at their respective positions.

EPA — This regulatory agency does not accept the classification of carcinogens into the categories of promoters and initiators. Accordingly, all suspected cancer-causing chemicals are assumed to act as initiators. This assumption requires a dose response relationship so that a curve can be drawn through the response points and extrapolated to zero risk and zero concentration (see Chapter 4 for a discussion of this technique). The data in Table 12 for male mice treated with UDMH will be examined (Figure 12). The problem is to fit a line through these points and extrapolate to zero. There are two ways of performing this task: (1) use the sophisticated Crump Global 86 computer program, or (2) visually draw a line of best fit through the two points including the origin and figure out the slope. Using the raw data, the Crump program estimated a slope of 0.0078 (mg/kg/day)$^{-1}$, the value that will be used in the remainder of this discussion.[25] Calculating the slope from the graph yielded 0.0075 (mg/kg/day)$^{-1}$.

Since the study was only for 1 year, a correction was necessary to convert the data to a lifetime, which in mice is 2 years. Tumor incidence has been empirically shown to increase by a power of 3 with time. Accordingly, in going from 1 to 2 years, the slope will be increased by a factor of 8[26] for a final potency of 0.0624 (mg/kg/day)$^{-1}$.

One final correction is necessary. Recall that humans are more sensitive by a factor of the cube root of the ratio of weights.[27]

$$(70/0.03)^{1/3} \text{ or } 13.26$$

Therefore, the potency or Q* for humans is 13.26 × 0.0624 or 0.83 (mg/kg/day)$^{-1}$. The account in the *Federal Register* (Moore, 1989) does not show how the EPA developed its potency factor. However, since the EPA's factor of 0.88 (mg/kg/day)$^{-1}$ is close to the above, a similar procedure must have been used. Once exposure is estimated, the risk is decided by multiplying the Q* by the dose or exposure. The EPA's value for Q* of 0.88 (mg/kg/day)$^{-1}$ will be used in the remainder of this discussion.[28]

The daily exposure to UDMH for adults is decided as follows. The marketbasket survey suggested a value of 0.228 mg/kg as the maximum contamination of UDMH found in processed apples. This will be combined with the observation that adults consume 14 g of products a day for a lifetime. A value for the exposure can be estimated by substituting these two pieces of information into the following relationship.

$$Exposure = \frac{0.228 \ mg \ UDMH}{kg \ apples} \times \frac{0.014 \ kg \ apples}{day \cdot adult} \times \frac{1}{70 \ kg/adult}$$

$$= 4.56 \times 10^{-5} \ mg/kg/day$$

The EPA must have used a different factor for daily consumption to arrive at

---

[25] The value for the Crump slope was determined by the program installed on a computer at the Dow Chemical Co., Midland, MI.

[26] [Lifetime of animal (2 years)/length of study (1 year)]^3 = (2/1)^3 or a factor of 8.

[27] Actually this relation converts the surface area of humans to the surface area of the test animal.

[28] It has since been learned that using the results of the full study the potency factor has been reduced from 0.88 to 0.46 (mg/kg/day)$^{-1}$ (Marshall, 1991).

its reported exposure of 5.1×10^{-5} mg/kg/day (Moore, 1989). The EPA's value will be used to decide the lifetime risk of humans exposed to UDMH.

$$\text{Risk} = Q^* \times \text{exposure}$$
$$= 0.88 \ (\text{mg/kg/day})^{-1} \times 5.1 \times 10^{-5} (\text{mg/kg/day})$$
$$= 4.5 \times 10^{-5}$$

This is interpreted as 4.5 additional cancer deaths in a population of 100,000 over a 70-year lifetime. To quote from the *Federal Register* (Moore, 1989)

". . . . The Agency believes that the lifetime risk to the U.S. population estimated from the available data (4.5×10^{-5}) are of sufficient concern that it should move as quickly as possible toward cancellation of daminozide. . . ."

As previously described, there are several flaws in this analysis. Three of the most important are as follows:

1. The response of 1 tumor out of 45 animals at the dose of 11.5 mg/kg/day is statistically insignificant and should not be used in a dose response curve.
2. The maximum dose used is at or close to the maximum tolerated dose as evidenced by the early deaths of the animals in the study (Moore, 1989).
3. Extrapolating a 1-year study to a full lifetime is a questionable practice. This needs to be discussed in a scientific forum before being used in a risk assessment study.

Natural Resources Defense Council — The NRDC report "Intolerable Risk: Pesticides in Our Children's Food" is seriously flawed. However, in spite of all these problems it caused a great deal of harm and therefore must be discussed. As might be anticipated, there have been many critiques (Shimskey, 1989, being but one example). One major criticism is that it is impossible to reconstruct how the risk assessment was found since the necessary data were not presented. For example, NRDC cites a cancer potency factor of 8.9 for UDMH and fails to give any units (NRDC, 1989). In addition the risk models have been modified from the ones found acceptable by EPA—and they have been altered to generate higher values. Whenever any mathematical model is used to "explain" cancer results or any other phenomena, they should have at least achieved some general acceptance from the scientific community.

In deciding residue levels for the exposure side of the equation, the NRDC felt that the levels should be higher than those given by the EPA. Accordingly, they decided to use the actual testing results from the FDA laboratories in

California. NRDC decided to throw out the data from the San Francisco laboratory and rely on the results from Los Angeles (Shimskey, 1989). The reported reason for this choice is that the L.A. laboratory had a much higher detection rate and, consequently, the values were higher. The results from the San Francisco laboratory were statistically insignificant, i.e., chemicals were not detected. Recent reports (Shimskey, 1989) showed that the L.A. laboratory tested produce predominantly from Mexico while the Bay area tested produce from the San Joaquin Valley, the major production area for the U.S. Thus, by selecting data the NRDC could inflate their risk estimations. As Henry Voss, director of the California Department of Food and Agriculture, said:

> ". . . it is a good example of what has been called 'political toxicology' which is motivated by social goals not scientific truth. The net result is that the public has been unduly alarmed."

Conclusion

The case for the chemical daminozide and products such as Alar has been concluded since the chemical is no longer being used. What remains is a much larger issue that is presently wending its way through the courts. A group of Washington state apple growers is using a class action lawsuit to seek damages as well as answers to a series of questions that should have been asked and answered long ago. The lawsuit names NRDC, CBS, and the public relations firm of Fenton Communications as the defendants. The complaint filed in Yakima County Superior Court in the State of Washington on November 28, 1990, lists three causes for action.

1. Product disparagement.
2. Tortious interference with business expectations.
3. Violation of Washington's unfair business practices.

The plaintiffs, according to their attorney Jay Sandlin (Stockwin, 1991):

> ". . . . are a group of family farmers growing apples in the state of Washington who never had a chance to be heard and who were injured when their apples were devalued as a result of false, misleading and scientifically unreliable statements made by the defendants."

The outcome will depend heavily on the facts brought by the plaintiffs to answer the following:

1. The defendants published a falsehood.
2. The defendants acted with reckless disregard of the truth or falsity of the statements made.
3. The falsehood was communicated to a third person.
4. The falsehood played a material and substantial part in inducing others not to deal with the plaintiffs.
5. The plaintiffs suffered special (financial) damages in the loss of trade or other dealings.

The defendants have already said they intend to base much of their defense on First Amendment guarantees of free speech. NRDC claims a constitutional right to publish its report that launched the media blitz as part of an ongoing scientific debate over public health. The plaintiff's attorney responded by stating:

> "... We're not against the scientific community having a debate.... How is that debate helped by using 10-second soundbites on national television to present a highly technical report that takes a minimum of three hours to read and a background in calculus, differential equations, toxicology, chemistry and statistics to make any sense of what you're reading? We intend to show not only that their study was the product of questionable science, but also that the reasons they published had nothing to do with furthering public debate."

The outcome from this trial will be most important. The Washington apple growers are interested in a settlement that would establish a standard of conduct for future cases when growers and other users of chemicals may be injured by falsehoods. A settlement in favor of the plaintiffs would cause high-profile media groups such as CBS to rethink their future course of action. For example, they might come to the conclusion that real science does not operate by leak or press conference. It is not propagated by movie stars. It can only be found in scientific journals where it has undergone the rigors of peer review. While such publications do not guarantee 100% accuracy, they do represent the best-researched scientific consensus at the time. This is the type of science that needs to be communicated to the public.

One final observation: this case is a classic example of the socioscientific issue referred to in Chapter 8. These are issues that are complicated in the scientific sense and are issues about which the public wants an answer. In the case of Alar, the public wants to know if apples are safe to eat. The strategy being used by the defendants appears to be delay tactics as opposed to the Rules of Reason (discussed in the previous chapter). The consumer would be best served if a serious effort could be made to arrive at some semblance of the truth and answer the question are children at risk by consuming apples contaminated with trace amounts of Alar?

APPENDIX I: STATISTICS

INTRODUCTION

The three topics of probability, expected value, and decision rules are important subjects in the study of statistics. However, they are not required for an understanding of the material in the text. They are included in this appendix to give the student an opportunity for further insight into the fascinating field of statistics.

PROBABILITY

Sampling was illustrated in Chapter 2 with a bag containing 100 white and 100 red beads. When samples are drawn from this population, i.e., drawing 10 beads from the bag of 100, the observations vary from sample to sample. The results of each trial cannot be predicted and are commonly thought of as being random or subject to chance. Although accurate predictions of a single experiment are impossible, a sequence of trials reveals a regularity. Referring to the bag of 100 beads, repeated trials show a convergence on 1.0 as the true ratio of red to white beads. In testing the trend it is customary to start with a hypothesis or a model to describe the system. Such a model is referred to as a probability model. In this context probability is the long-term frequency that the event under investigation will occur. In the case of the beads the model is that the ratio of red to white is 1. By drawing samples the model is tested by establishing that the long-term frequency of the ratio is indeed 1. Probability is expressed as a decimal fraction between zero (the event never occurs) and unity (the event always occurs). Sometimes a probability is known *a priori*, as in penny tossing, where in the long run heads and tails fall with a frequency of 0.5 which is the probability of a single toss producing a head or a tail. In other problems, a probability can only be estimated empirically by observing the relative frequency to which the results of a great many trials converge. These two types of probabilities and their associated models are

Table 1.   Possible Outcomes from Throwing Two dice

1, 1	2, 1	3, 1	4, 1	5, 1	6, 1
1, 2	2, 2	3, 2	4, 2	5, 2	6, 2
1, 3	2, 3	3, 3	4, 3	5, 3	6, 3
1, 4	2, 4	3, 4	4, 4	5, 4	6, 4
1, 5	2, 5	3, 5	4, 5	5, 5	6, 5
1, 6	2, 6	3, 6	4, 6	5, 6	6, 6

referred to as deterministic and stochastic, respectively. The use of stochastic models is reserved for events that are too complicated to establish a deterministic model. For example, the weatherman uses a series of empirical relations to predict weather. This is necessary because forecasting the weather from a mechanistic point of view is impossible. In many instances the stochastic model is the only kind that has any practical value.

In the case of the deterministic model a definite mathematical structure can be assigned that will accurately predict the long-term outcome of an event. Penny tossing is one example; throwing two dice is another. In this latter case, the outcome of all possible experiments (a total of 36) is shown in Table 1. By using this table a sampling distribution can be established showing the frequencies of obtaining various values. For example, the model would indicate that there are 11 discrete sums. The value 2 or 12 will appear 1/36 times which translates to a probability of 0.03. The remaining sums are shown in the histogram in Figure 1. Using this model questions regarding possible outcomes in throwing dice can be readily answered. Thus, the probability of throwing a 4 or less is the sum of the probabilities for 4, 3, and 2 or 0.08 + 0.06 + 0.03 = 0.17. Be assured that the operators of the casinos in Reno and Las Vegas are well aware of all these probabilities.

EXPECTED VALUE

The previous discussion of probabilities and throwing dice leads naturally to the concept of expected value. When dice are thrown the long-term average result is called the expected value. The following example (Lindgren and McElrath, 1971) illustrates the idea more clearly. In a certain gambling house a game consists of a toss of a coin with a $1 payoff for heads and a $2 payoff for tails. If the game is to be exactly fair how much should be charged to play the game?

In a large number (N) of tosses the number of heads is about N/2 and the number of tails is N/2. In those N plays of the game the house will pay $1 for N/2 heads and $2 for N/2 tails. The total paid out is

$$\text{Total} = \$1 \times N/2 + \$2 \times N/2 = \$3 \times N/2$$

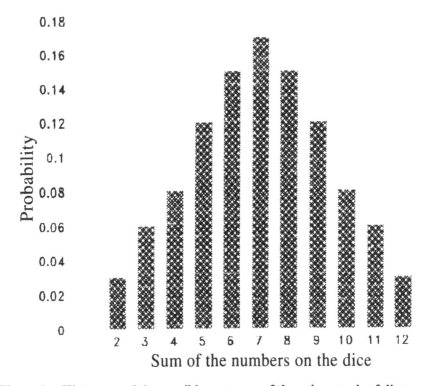

Figure 1.   Histogram of the possible outcomes of throwing a pair of dice.

The house needs to collect $3N/2 from the N players in order to break even. This works out to be $1.50 for each player. Such a payment is called the expected value or sometimes the mathematical expectation, the adjective "mathematical" implying that the expectation is an idealization as N becomes large without limit and the proportion converges to probabilities. Note that the expected value, a payout of $1.50, never occurs; it is simply the long-term average.

Another problem to illustrate both probability and expected value involves a one-and-one free-throw shooter in a basketball game.[1] "The home town team" is behind by one point. As the buzzer sounds Terry is fouled and goes to the line for a one-and-one foul shot. Terry is a 60% "free-throw shooter" What is the expected value for scoring in this situation? Would you bet that he would win the game? What odds would you give? Using an area model illustrates the problem and solution. Divide a 10 × 10 grid in the ratio of 60:40 (Figure 2a). This represents 60% of the time that Terry hits the first shot and has an opportunity to try again. By dividing the hit area in Figure

---

[1] This example was taken from a teaching aid published in "Natural Science," College of Natural Science, Michigan State University, East Lansing, MI, 1989.

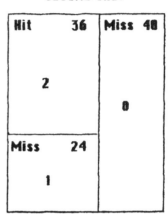

FIRST SHOT

| Hit | 60 | Miss 40 |

SECOND SHOT

Hit	36	Miss 40
	2	0
Miss	24	
	1	

a   b

Figure 2.   Grid illustrating the solution ot the 60% free-throw shooter. (a) First shot and (b) the results of the second shot.

2b for the second shot into 60:40, the squares can be counted to arrive at a solution.

$$P(0) = 40/100, P(1) = 24/100, \text{ and } P(2) = 36/100$$

where

$P(x)$ represents the probability of event x happening

Another solution utilizes the theoretical construct that the total probability of an event is estimated by multiplying the individual probabilities. For example, the probability of making only 1 point is the probability of making the first shot multiplied by the probability of missing the second.

$$P(1) = 60/100 \times 40/100 = 0.24$$

The assumption is that every shot has a 0.6 probability of scoring, regardless of the outcome of the previous shot.

In a similar manner

$$P(2) = 60/100 \times 60/100 = 0.36$$

and finally

$$P(0) = 40/100 = 0.4$$

Notice that the total probabilities add up to 1. In other words, the probability of 0, 1, and 2 points is a certainty. If the expected value is the average number of points scored over many trials, what is the expected value for Terry? In 100 trips

$$P(0) = 0.4$$
$$P(1) = 0.24$$
$$P(2) = 0.36$$
$$\text{Total points} = (0.4 \times 0) + (0.24 \times 1) + (0.36 \times 2) = 0.96$$

The expected value is 0.96 per trip for the free throw. Conclusion: if you bet that Terry will win the game, make sure you have odds in your favor!

DECISION RULES

In the section on sampling (Chapter 2) the concept of decision making and hypothesis testing was introduced. A hypothesis is a statement of what the sample represents. For example, in the case of the eggshell experiment discussed in Chapter 2 the hypothesis is that the egg was a member of the control population. By measuring the thickness of the eggshell and comparing the statistics, a decision was made regarding the validity of the hypothesis. If the hypothesis was rejected, then the alternate hypothesis must be accepted. Again in the case of the eggshells the alternate hypothesis is that the eggshell thickness is affected by the treatment. Thus, every time there is a hypothesis there must be an alternate hypothesis. At what point is the hypothesis rejected? The answer will be illustrated using the following scenario. Imagine a penny that shows a head and a decision must be made regarding the hidden side. In other words, does the penny have two heads or a tail and a head? To make the decision the penny can be tossed as many times as needed. The best way to proceed is to toss the coin. If a tail appears, the problem is solved. If a tail does not appear over a sequence of trials, a decision must be made at some point that the penny is two-headed. No matter when the decision is made you may be wrong; the very next toss might be a tail. The longer you toss, the more certain that the correct decision must be that the penny is two-headed. The problem becomes more interesting and more realistic when a cost is attached to each toss and a penalty is associated with a wrong decision. If the cost of continued tossing was low and the penalty was high for a wrong decision, then you would like to see a long succession of heads before deciding. On the other hand, if the cost of tossing was high and the penalty for a wrong decision was low, the decision to terminate would be much earlier. In order to make the decision a hypothesis needs to be formulated. In the case of the coin the hypothesis might be that the coin has both a head and a tail. Once the model is established the next decision deals with how often should the hypothesis be rejected when in fact it is correct. Again using the case of the coin the question needs to be asked how often should the model of a true coin be rejected when in fact the coin does have both a head and a tail? In the vernacular of statistics this is called a type I error and the probability of committing a type I error is the same as the level of significance. If the type I error is set at 5%, then 5 out of every 100 times the experiment is repeated the hypothesis will be rejected when it is true. Type II error is the reverse in that the hypothesis is accepted when it is false. As the chance of rejecting a true hypothesis is made small ($p = 0.001$), then the corresponding type II error becomes large. In other words, the hypothesis that the coin is normal will be accepted even though the odds are increasing dramatically that the coin is indeed false. The only time the type I error is set very small is if the penalty for accepting the alternate hypothesis is very costly.

Returning to the coin toss, the probabilities for a series of trials is as follows:

Probability of 1 head		$= 0.5$
Probability of 2 successive heads	$= 0.5 \times 0.5$	$= 0.25$
Probability of 3 successive heads	$= 0.5^3$	$= 0.13$
Probability of 4 successive heads	$= 0.5^4$	$= 0.06$
Probability of 5 successive heads	$= 0.5^5$	$= 0.03$

Thus, the probability of having 4 heads in a row is 0.06. This exceeds the type I error of 0.05 and thus the hypothesis that the coin is normal cannot be rejected. However, the probability of seeing 5 heads in a row is 0.03 which is smaller than the previously set level of significance. In this case the hypothesis is rejected and we accept the fact that the coin is two-headed. The decision rule for this game is "accept the hypothesis if any tail appears but reject the hypothesis if no tail appears within 5 tosses." For two-headed coins this rule is free of error. For honest coins there is a 3 out of 100 chance that an incorrect decision will be reached.

In biological work an arbitrary tradition is to adopt $p < 0.05$ as a reasonable compromise between type I and type II errors. This means that only once in 20 times will an incorrect decision of a significant difference be reached. If p is set at 0.01 then the conclusions are "highly significant," i.e., there is only 1 chance in 100 that they will be wrong. These conventions are arbitrary but are normally sanctioned by custom. In the case of the DDT experiment (Bitman et al., 1969) the results of the treatment were reported as highly significant with a p of 0.01. Thus, the results of adding DDT to the diet of birds caused a statistically significant reduction in the thickness of eggshells. Finally, the difference between "statistical significance" as expressed by the p values and "practical significance" needs to be mentioned. For example, a drug might be developed which lowers body weight by a statistically significant amount and still be insignificant from an overall weight reduction program. The experiments cited in Chapter 2 on DDT and eggshell thickness did not address the question of practical significance. However, there is no doubt from many other experiments that DDT in the diet of birds has a significant impact on lowering the reproductive rate (Carson, 1962).

APPENDIX II: MATHEMATICS

EXPONENTIAL CURVES

Whenever the amount of a product increases with time at a constant rate, the growth follows an exponential curve. Many environmental problems follow this pattern. For example, the increasing use of pesticides such as DDT, the consumption of volatile halocarbons such as the Freons, and the production of industrial materials such as polychlorinated biphenyls (PCBs) fit the picture of exponential growth. Consequently, these curves need to be understood.

There are many examples of this type of growth. A bacterial cell that divides every 20 minutes is growing exponentially. After 20 minutes there will be 2 cells, after the next 20 minutes there will be 4 cells, then 8, 16, etc. In a similar fashion if a person invests $10.00 at 7%, exponential growth in the amount of money results. At the end of 30 years the total would be $80.00 (three doubling times). On the other hand, if the $10.00 grew in a linear fashion, then every year would add $0.70 and at the end of 30 years there would be the original $10.00 plus $21.00 in growth for a total of only $31.00. Exponential growth is a common type of curve and there are many illustrations of the surprising consequences. There is an old Persian legend about a clever person who presented the king with an ornate chessboard and requested the king give him in return 1 grain of rice for the first square, 2 for the second and so forth for all 64 squares. The king readily agreed and ordered rice from the storehouse. By the 40th square a trillion grains were required. As you might imagine the king's entire rice crop was exhausted before reaching the 64th square. Exponential increase is deceptive because huge numbers are generated very quickly.

A useful way of thinking about exponential growth is in terms of doubling times, or the time it takes for an item to double in size. A sum of money left in the bank at 3% doubles in about 23 years. Equation 1 shows a simple relation between the rate of growth and the time for doubling.

$$\text{Doubling time} = 70/\text{growth rate in \%} \tag{1}$$

The differential equation for this type of curve is given in Equation 2 where the rate of growth (dA/dt) is given as a function of time.

$$dA/dt = kt \tag{2}$$

Integration of Equation 2 yields Equation 3, which is the mathematical description of an exponential equation where the amount of A is growing at a constant rate represented by the rate constant k.

$$A = A_0 \exp(kt) \tag{3}$$

where

$$A_0 = \text{initial amount}$$

Setting $A = 2A_0$ (doubling the initial amount) yields the relation shown in Equation 4.

$$t = (\ln 2)/k \tag{4}$$

Since the value of the natural logarithm of 2 is 0.69, the doubling relation mentioned earlier is obtained.[1]

In a similar fashion the time for an item to decrease by half can be illustrated. This is referred to as the half-life or the length of time to decrease the original amount by one half. In this case the rate constant k has a negative sign and the differential equation is given by Equation 5:

$$dA/dt = -kt \tag{5}$$

Integration of the equation yields

$$A = A_0 \exp(-kt) \tag{6}$$

By setting $A = 1/2\ A_0$ (decreasing the original by half) the relation shown in Equation 7 is obtained.

$$t_{1/2} = (\ln 0.5)/-k \tag{7}$$
$$= 0.69/k \text{ (identical to doubling time)}$$

Notice that the time in this case is called the half-life and is represented by $t_{1/2}$.

---

[1] Multiplying Equation 4 by 100 changes k to % and the 0.69 to approximately 70.

MASS OF ATMOSPHERE AND CURVE FITTING

The density of the atmosphere as a function of altitude is shown in Table 1 (Ebbing, 1987). The data in this table form an exponential curve where the density decreases with increasing altitude at a fixed rate. To a first approximation the relation between density and altitude can be fitted to Equation 8.

$$d = d_0 \exp(-Kh) \tag{8}$$

where

$K = 1.069 \times 10^{-4} \text{ m}^{-1}$ (curve fitting constant)
$h =$ height in meters
$d_0 =$ density at sea level (g/m^3)

From the relation in Equation 8 the weight of the atmosphere is given by Equation 9.

$$Weight = d_0\, A \int \exp(-Kh)\, dh$$
$$A = \text{surface area of the earth in m}^2 \tag{9}$$
$$= 5.10 \times 10^8 \text{ km}^2 \text{ (Weast, 1986)}$$

Integration of Equation 9 yields 4×10^{21} grams as the mass of the troposphere. The troposphere is defined in this calculation as the lower 10 km of the atmosphere. The *CRC Handbook of Chemistry and Physics* (Weast, 1986) cites a value of 5.2×10^{21} g for the mass of air. One possible reason for the discrepancy in the two numbers may be how much atmosphere was included in the two estimations.

MASS BALANCE CALCULATION OF FREONS IN THE TROPOSPHERE

There are many types of mass balance studies ranging from the very complex to the very simple. The goal in all cases is to match the materials

Table 1. Density of the Atmosphere as a Function of Altitude

Altitude (km)	Density (g/m³)
0	1220
4	803
6	649
8	524
10	419
20	92

that enter a system with the materials that exit. In essence they are accounting procedures. The following relationship is useful for understanding the process.

$$(\text{Total material}) = (\text{material added} - \text{material lost}) \tag{10}$$

This implies that for a closed system the total mass is constant. Furthermore, at equilibrium the rate of change of the mass in the various compartments is zero. Once the equations for the system have been developed, the mass relations should be examined very carefully. This will prevent the embarrassing situation where the conservation law has been violated and the system is found to be creating mass without an appropriate source term.

An elementary mass balance approach will be presented for the Freons. The procedure is based on the assumption that all the material ends in the atmosphere. This is a reasonable assumption since the agents are volatile and are used in such a manner that they are free to enter the atmosphere. Thus, by knowing the volume of the troposphere and the mass of Freons that have been produced, an estimation of the concentration may be made. This can be matched with analytical data and a decision made as to how much of the original material is still present and how much has been degraded or lost. The following represents a very simple mass balance study and illustrates the type of accounting that is involved.

1. The volume of the troposphere is estimated assuming a height of 10 km and knowing the radius of the earth (6371 km)

$$V = 4/3 \ \pi[(r + 10)^3 - r^3] = 5.1 \times 10^9 \ km^3$$

2. The mass of air in this volume is simply the volume times the density or 4.1×10^{21} g. From Table 2 in Chapter 5 an average density for air is about 800 g/m^3. This is very close to the value estimated in the previous section where the density altitude data was fitted to a curve.

3. The molecular weight of air is approximately equal to the weight of nitrogen, or 29.

4. The number of moles of air in the troposphere is 1.4×10^{20} (mass of air/molecular weight of air).

With these facts the following "back of the envelope" calculations for CCl_3F can be made. Production grew at a rate of about 18% per year for 17 years. The following formula can be used to estimate the total mass added to the atmospheric compartment.

$$\Sigma = A \ (r^y - 1)/(r - 1)$$

where

$$A = \text{initial production} \ (2.97 \times 10^{10} \ \text{g/year})$$
$$r = 18\% \ \text{or} \ 1.18$$
$$y = 17 \ \text{years}$$
$$\Sigma = 2.59 \times 10^{12} \ \text{g}$$

The molecular weight of CCl_3F is 137, yielding $2.59 \times 10^{12}/137$ or 1.89×10^{10} mol. This represents the total production over the 17-year lifespan.

From analytical monitoring studies the concentration of CCl_3F in the troposphere was 126 pmol/mol or 126 mol/trillion moles of air (NAS, 1976). The number of moles of air in the troposphere from the prior estimation was 1.4×10^{20} mol. Consequently, the number of moles of CCl_3F released into the atmosphere was calculated as follows:

$$\frac{126 \ \text{mol of CFC}}{10^{12} \ \text{mol of air}} \times \frac{1.4 \times 10^{20} \ \text{mol of air}}{\text{troposphere}}$$

$$\text{or} \ 1.76 \times 10^{10} \ \text{mol of} \ CCl_3F$$

This amount compares favorably to the 1.89×10^{10} mol from production figures.

These were the types of calculations that led people to the following understanding: since the Freons were not destroyed in the lower atmosphere, they would eventually have to go somewhere. The most probable location was a transfer to the stratosphere. This was the hypothesis that led investigators to examine the fate of chlorofluorocarbons and other halocarbons such as methylchloroform in the upper atmosphere.

REFERENCES

ACGIH. 1975. *Threshold Limit Values.* Washington, DC: American Conference of Governmental and Industrial Hygienists.

Beeton, A. M. 1979. *Polychlorinated Biphenyls.* Washington, DC: National Academy of Sciences.

Bitman, J., H. C. Ceal, S. J. Harris, and G. F. Fries. 1969. *Nature* 224:44.

Bolin, B. 1970. *Sci. Am.* 223:130.

Boyle, M. 1988. *Environ. Sci. Technol.* 22:1397.

Brand, D. 1988. Is the earth warming up? *Time* July 4, p. 18.

Brodie, B. B. 1964. *Pharmacologic Techniques in Drug Evaluation.* Nordine, J. H., and P. E. Siegler, Eds., p. 69. Chicago: Yearbook Medical Publishers.

Bumb, R. R., W. B. Crummett, S. S. Cutie, J. R. Gledhill, R. H. Hummel, R. O. Kagel, L. L. Lamparski, E. V. Luoma, D. L. Miller, T. J. Nestrick, L. A. Shadoff, R. H. Stehl, and J. S. Woods. 1980. *Science* 210:385.

Bureau of the Census. 1982. Statistical Abstracts of the United States. Washington, DC: U.S. Department of Commerce.

Carson, R. 1962. *Silent Spring.* Boston: Houghton Mifflin.

Cohen, B. 1986. *Consumers Res.* April, p. 11.

Costle, D. 1980. *Fed. Reg.* 45:79318, Nov. 28.

Croyle, R. M. 1989. Personal communication from the Dow Chemical Company, Midland, MI.

Ebbing, D. O. 1987. *General Chemistry.* p. 452. Boston: Houghton Mifflin.

Elmer-Dewitt, P. 1992. Summit to save the earth. *Time* June 1, p. 42.

Ember, L. R., P. L. Layman, W. Lepkowski, and P. S. Zurer. 1986. Changing atmosphere: implications for mankind. *Chem. Eng. News* Nov. 24, p. 14.

EPA. 1983. Dow Chemical Co.—Midland Plant Wastewater Characterization Study. March 23. Chicago: U.S. EPA Region V.

EPA. 1986. A Citizen's Guide to Radon. August, EPA-86-004.

Exton, C. 1989. Data package provided by Uniroyal Chemical Co., Middlebury, CT.

FDA. 1979. *Fed. Reg.* 44:38330, June 24.

Fisher, A. C., and W. Worth. 1988. *Cancer in the United States—Is There an Explosion?* New York: American Council on Science and Health.

Gough, M. 1986. *Dioxin, Agent Orange: The Facts.* New York: Plenum Press.

Hammond, A. L. 1977. *Science* 195:164.

Hansen, D. J. 1987. Agent Orange. *Chem. Eng. News* Nov. 15, p. 7.

Hansen, D. J. 1989. Radon tagged as cancer hazard by most studies. *Chem. Eng. News* Feb. 6, p. 7.

Hardin, G. 1968. *Science* 162:1243.

Higgison, J. 1969. *Proc. 8th Canadian Cancer Research Conference.* p. 40. New York: Pergamon Press.

Hileman, B. 1989. Global warming. *Chem. Eng. News* March 13, p. 25.

Kaczmar, S. W., M. J. Zabik, and F. M. D'Itri. 1983. *American Chemical Society,* 186th National Meeting, 23:84. Washington DC.

Kahneman, D., P. Slovic, and A. Tversky. Eds. 1982. *Judgement under Uncertainty: Heuristics and Biases.* New York: Cambridge University Press.

Kahneman, D., and A. Tversky. 1973. *Psychol. Rev.* 80:237.

Kociba, R. J., D. G. Keyes, J. E. Beyer, R. M. Carreon, C. E. Wade, D. A. Dittenbar, R. P. Kalnis, L. E. Frauson, C. N. Parks, S. D. Barnard, R. A. Hummel, C. G. Humiston. 1978. *Toxicol. Appl. Pharmacol.* 46:279.

Lamparski, L. L., T. J. Nestrick, and R. H. Stehl. 1979. *Anal. Chem.* 51:1453.

Lemonick, M. D. 1989. Feeling the heat. *Time* Jan. 2.

Linden, E. 1989. Playing with fire. *Time* Sept. 18, p. 76.

Lindgren, B. W., and G. W. McElrath. 1971. *Introduction to Probability and Statistics.* 3rd ed. New York: Macmillan.

Loomis, T. A. 1978. *Essentials of Toxicology.* Philadelphia: Lea & Febiger.

Lovelock, J. E. 1971. *Nature* 230:379.

Lovelock, J. E. 1977. *Nature* 267:32.

Mackay, D. 1991. *Multimedia Environmental Models.* Boca Raton, FL: Lewis Publishers.

Mackay, D., S. Paterson, S. J. Eisenreich, and M. S. Simmons. 1983. *Behavior of PCBs in the Great Lakes.* Ann Arbor, MI: Ann Arbor Science.

Marshall, E. 1991. *Science* 254:20.

McKean, K. 1985. Decisions. *Discover* June, p. 22.

Miller, G. T. 1985. *Living in the Environment.* p. 367. Belmont, CA: Wadsworth Publishing.

Molina, M. J., and F. S. Rowland. 1974. *Nature* 249:810.

Moore, D. S. 1990. *On the Shoulders of Giants—New Approaches to Numeracy.* Steen, A. L., Ed. Washington, DC: National Academy Press.

Moore, J. A. 1987. *Fed. Reg.* 52:28256, July 29.

Moore, J. A. 1989. *Fed. Reg.* 54:6392, Feb. 10.

Morgan, M. G. 1985. *Environmental Exposure from Chemicals.* Vol. 2. Neely, W. B., and G. E. Blau, Eds., p. 107. Boca Raton, FL: CRC Press.

Nader, R. 1965. *Unsafe at Any Speed.* New York: Grossman Publishers.

NAS. 1979. *Stratospheric Ozone Depletion by Halocarbons: Chemistry and Transport.* Washington, DC: National Academy of Sciences.

NAS. 1982. *Causes and Effects of Stratospheric Ozone Reduction: An Update.* Washington, DC: National Academy of Sciences.

NAS. 1983. *Risk Assessment in the Federal Government: Managing the Process.* Washington, DC: National Academy of Sciences.

NAS. 1976. *Halocarbons: Environmental Effects of Chlorofluoromethane Release.* Washington, DC: National Academy of Sciences.

Neely, W. B. 1977. *Sci. Total Environ.* 8:267.

Neely, W. B. 1992. *Emergency Response from Chemical Spills.* Boca Raton, FL: Lewis Publishers.

Neely, W. B., and J. H. Plonka. 1978. *Environ. Sci. Technol.* 12:317.

Neely, W. B. 1980. *Chemicals in the Environment.* New York: Marcel Dekker.

Neely, W. B. 1985. In *New Approaches in Toxicity Testing and Their Application in Human Risk Assessment.* Li, A. P., Ed. p. 235. New York: Raven Press.

Neely, W. B., and G. R. Oliver. 1986. In *Pollutants in a Multimedia Environment.* Cohen, Y., Ed. p.133. New York: Plenum Press.

Neely, W. B. 1985. *Aquatic Toxicology and Hazard Assessment.* Bahner, R. C., and D. J. Hansen, Eds. p. 455. (8th Symposium), ASTM STP 891, Philadelphia: American Society for Testing and Materials.

Nero, A. V. 1992. *Issues* Fall, p. 23.

NIOSH. 1980. Registry of Toxic Effects of Chemical Substances. Washington, DC: U.S. Department of Health and Human Services (1991 Microfiche update was examined for the most recent results).

NRC. 1988. *Health Risks of Radon and Other Internally Deposited Alpha Emitters.* BEIR IV, Committee on the Biological Effects and Longer Radiation. Washington, DC: National Academy Press.

NRDC. 1989. Intolerable Risk: Pesticides in Our Children's Food. Report by the Natural Resources Defense Council, Washington, DC.

Park, C. N., and R. D. Snee. 1983. *Am. Statistician,* 37:427.

Prescott, D. M., and A. S. Flexer. 1982. *Cancer, the Misguided Cell.* New York: Charles Scribner's Sons.

Ratcliffe, D. A. 1967. *Nature* 215:208.

Revkin, A. C. 1988. Endless summer: living with the greenhouse effect. *Discover* October, p. 50.

Rogers, E. M. 1989. Life in the Balance. Dow Chemical Co., Midland, MI.

Rowland, F. S. 1989. *Am. Sci.* 77:36.

Santerre, C. R., J. N. Cash, and M. J. Zabik. 1991. *J. Food Prod.* 54:225.

Saxton, W. L., K. Steinbrecher, E. Gunderson. 1989. *J. Ag. Food Chem.* 37:570.

Shimskey, D. S. 1989. *Am. Fruit Grower* May.

Simpson, C. F., and M. V. Barrow. 1972. *Arch. Environ. Health* 25:349.

St. Johns, L. E., H. Arnold, and D. J. Liske. 1969. *J. Ag. Food Chem.* 17:116.

Stockwin. 1991. *Am. Fruit Grower* January.

Strahler, A. N., and A. H. Strahler. 1973. *Environmental Geoscience: Interaction Between Natural Systems and Man.* p. 289. Santa Barbara, CA: Hamilton Publishing.

Toth, B., L. Wallcave, K. Patil, I. Schmeltz, and D. Hoffmann. 1977. *Cancer Res.* 37:3497.

Tshirley, F. H. 1986. *Sci. Am.* 254:29.

Turner, B. D. 1970. Workbook of Atmospheric Dispersion Estimates. PB-191 482, Cincinnati: National Air Pollution Control Administration.

USDA. 1985. Continuing Survey of Food Intake by Individuals—Nationwide Food Consumption Survey, Report 85-4, Washington, DC: U.S. Department of Agriculture.

Weast, R. C. 1986. *CRC Handbook of Chemistry and Physics.* 66th ed. Boca Raton, FL: CRC Press.

Wessel, M. R. 1980. *Science and Conscience.* New York: Columbia University Press.

Zimmer, C. 1993. The case of the missing carbon. *Discover* December, p. 38.

Zurer, P. S. 1989. CFC alternatives face phaseout. *Chem. Eng. News* Dec. 4, p. 9.

Zurer, P. S. 1988. Ozone depletion. *Chem. Eng. News* May 30, p. 16.

Zurer, P. S. 1992. Industry, consumers prepare for compliance with pending CFC ban. *Chem. Eng. News* June 23, p. 7.

Zurer, P. S. 1987. Antarctica ozone hole. *Chem. Eng. News* Nov. 2, p. 22.

Zurer, P. S. 1993. Looming ban on production of CFCs, halons spurs switch to substitutes. *Chem. Eng. News* Nov. 15, p. 12.

INDEX

Printed and bound by CPI Group (UK) Ltd, Croydon, CR0 4YY

24/10/2024

01778287-0001